Sydney Russo

STATES
ACADEMIC PRESS
www.statesacademicpress.com

Published by States Academic Press,
109 South 5th Street,
Brooklyn, NY 11249, USA

ISBN: 978-1-63989-230-3

Cataloging-in-Publication Data

Fundamentals of polymer chemistry / Sydney Russo.
p. cm.
Includes bibliographical references and index.
ISBN 978-1-63989-230-3
1. Polymers. 2. Polymerization. I. Russo, Sydney.
QD381 .F86 2022
547.7--dc23

For information on all States Academic Press publications visit our website at www.statesacademicpress.com

STATES
ACADEMIC PRESS

Contents

Preface

Polymers are large chains of molecules that consist of many repeating sub-units called monomers. On the basis of their structure, these can be classified into linear polymers, branched chain polymers and cross-linked polymers. The branch of chemistry which deals with the physical and chemical properties of polymers and macromolecules is called polymer chemistry. It attempts to study the various techniques and methods involved in the formation of polymers. Some of the major theories which serve as the guiding principles of polymer chemistry are Scheutjens-Fleer theory, Flory-Huggins solution theory, polymer field theory and Hoffman nucleation theory. Cellulose, chitin, starch, DNA and RNA are a few examples of naturally occurring polymers. Synthetic polymers include various thermosetting and thermoplastic polymers such as Bakelite, Kevlar, Teflon, polystyrene, polyethylene, etc. This book is a compilation of chapters that discuss the most vital concepts in the field of polymer chemistry. The various sub-fields of polymer chemistry along with technological progress that have future implications are glanced at herein. Those in search of information to further their knowledge will be greatly assisted by this book.

To facilitate a deeper understanding of the contents of this book a short introduction of every chapter is written below:

Chapter 1- The sub-discipline of chemistry that deals with the physical and chemical properties such as chemical synthesis and structure of polymers and macromolecules is known as polymer chemistry. A polymer is a substance which is composed of repeating subunits. Polymers are classified on the basis of their availability, structure of the monomer chain, polymerization, monomers and molecular forces. This is an introductory chapter which will introduce briefly all the significant aspects of polymer chemistry.

Chapter 2- The properties of polymers depend on their structure. Mechanical properties of polymers dictate how polymers behave on a macroscopic scale. Transport properties describe how rapidly molecules move through the polymer matrix. Various properties of polymers like molecular properties, transport properties, mechanical properties, optical properties, electrical properties and phase behavior etc. are discussed in detail in this chapter.

Chapter 3- Organic polymers are the polymers which contain carbon atoms in the backbone. Organic polymers play a crucial role in living organisms like providing basic structural materials and participating in vital life processes. This chapter has been carefully written to provide an easy understanding of the varied facets of organic polymers.

Chapter 4- Inorganic polymers are the polymers that do not contain carbon atoms in their backbone. They have ionic and hydrocarbon bonds in their backbone along with some heteroatom such as oxygen, nitrogen and sulphur. This chapter closely examines the key concepts of inorganic polymers to provide an extensive understanding of the subject.

Chapter 5- Polymer science is a subfield of materials science that involves polymers, mainly synthetic polymers like plastics and elastomers. Polymer architecture describes the shape of a single polymer molecule. Science and technology have undergone rapid developments in the past decade which has resulted in the discovery of significant tools and techniques in the field of polymer science and architecture which have been extensively detailed in this chapter.

Chapter 6- Flory-Stockmayer theory predicts the gel point of the system containing three types of monomer units; linear units with two A-groups, linear units with two B groups and branched A units. Flory-Huggins solution theory is the most prominent method for understanding phase behavior of polymer mixtures and thermodynamics. This chapter discusses in detail the theories and methodologies related to polymer chemistry.

Chapter 7- The chemical reaction in which monomer molecules react together to form polymer chains is known as polymerization. In radical polymerization method of polymerization, the polymer forms by the successive addition of free radical blocks. This chapter explains briefly various types of polymerization like step-growth polymerization, chain-growth polymerization, radical polymerization and copolymerization.

Chapter 8- Any change in the physical or chemical properties of a polymer or polymer-based products due to heat, light, chemicals or any other outer force. There are various processes of polymer degradation like UV degradation, Thermal degradation, thermal depolymerization and photo-oxidation of polymers. The topics elaborated in this chapter will help in gaining a better perspective about degradation of polymers.

I owe the completion of this book to the never-ending support of my family, who supported me throughout the project.

Sydney Russo

1

Polymer Chemistry: An Introduction

The sub-discipline of chemistry that deals with the physical and chemical properties such as chemical synthesis and structure of polymers and macromolecules is known as polymer chemistry. A polymer is a substance which is composed of repeating subunits. Polymers are classified on the basis of their availability, structure of the monomer chain, polymerization, monomers and molecular forces. This is an introductory chapter which will introduce briefly all the significant aspects of polymer chemistry.

Polymer chemistry is the study of the synthesis, characterization and properties of polymer molecules or macromolecules, which are large molecules, composed of repeating chemical subunits known as monomers. Sometimes this area of science is also called macromolecular chemistry. Essentially, careers in this subsector revolve around chemical synthesis (i.e. making chemicals artificially) and understanding the structure of these chemical chains. These chains can be either synthetic (manmade) or naturally occurring. Naturally occurring polymers like starch or protein are called biopolymers. Synthetic polymers are often just referred to as polymers and include common materials such as nylon or polyester. Polymer chemistry is a major contributor to advances in industry and medicine. A large number of everyday objects have been developed directly as a result of advances in polymer chemistry, such as plastic bags, disposable contact lenses or tyres.

Polymer

A polymer is a large molecule made up of chains or rings of linked repeating subunits, which are called monomers. Polymers usually have high melting and boiling points. Because the molecules consist of many monomers, polymers tend to have high molecular masses.

The word polymer comes from the *poly-*, which means "many," and the suffix *-mer*, which means "parts." The word was coined by Swedish chemist Jons Jacob Berzelius in 1833, although with a slightly different meaning from the modern definition. The modern understanding of polymers as macromolecules was proposed by German organic chemist Hermann Staudinger in 1920.

Examples of Polymers

Polymers may be divided into two categories. Natural polymers (also called biopolymers) include silk, rubber, cellulose, wool, amber, keratin, collagen, starch, DNA, and shellac. Biopolymers serve key functions in organisms, acting as structural proteins, functional proteins, nucleic acids, structural polysaccharides, and energy storage molecules.

Synthetic polymers are prepared by a chemical reaction, often in a lab. Examples of synthetic polymers include PVC (polyvinyl chloride), polystyrene, synthetic rubber, silicone, polyethylene, neoprene, and nylon. Synthetic polymers are used to make plastics, adhesives, paints, mechanical parts, and many common objects.

Synthetic polymers may be grouped into two categories. Thermoset plastics are made from a liquid or soft solid substance that can be irreversibly changed into an insoluble polymer by curing using heat or radiation. Thermoset plastics tend to be rigid and have high molecular weights. The plastic stays out of shape when deformed and typically decompose before they melt. Examples of thermoset plastics include epoxy, polyester, acrylic resins, polyurethanes, and vinyl esters. Bakelite, Kevlar, and vulcanized rubber are also thermoset plastics.

Thermoplastic polymers or thermosoftening plastics are the other type of synthetic polymers. While thermoset plastics are rigid, thermoplastic polymers are solid when cool, but are pliable and can be molded above a certain temperature. While thermoset plastics form irreversible chemical bonds when cured, the bonding in thermoplastics weakens with temperature. Unlike thermosets, which decompose rather than melt, thermoplastics melt into a liquid upon heating. Examples of thermoplastics include acrylic, nylon, Teflon, polypropylene, polycarbonate, ABS, and polyethylene.

Uses of Polymers

Polymers are used in almost every area of modern living. Grocery bags, soda and water bottles, textile fibers, phones, computers, food packaging, auto parts, and toys all contain polymers.

Even more-sophisticated technology uses polymers. For example, "the membranes for water desalination, carriers used in controlled drug release and biopolymers for tissue engineering all use polymers,". Popular polymers for manufacturing include polyethylene and polypropylene. Their molecules can consist of 10,000 to 200,000 monomers.

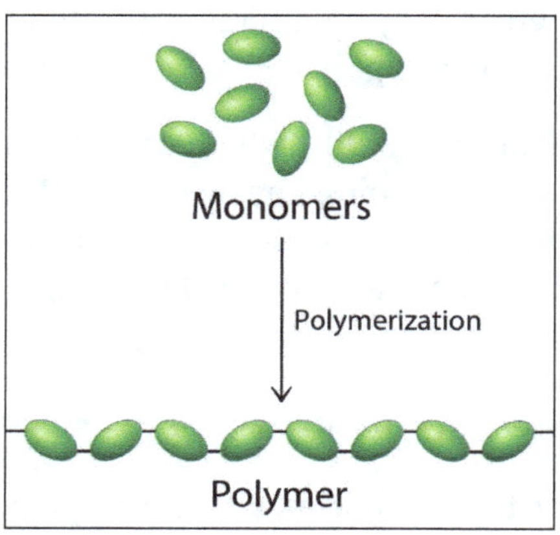

Figure: During a polymerization reaction, a large number of monomers become connected by covalent bonds to form a single long molecule, a polymer.

Biopolymers

Biopolymers are polymeric biomolecules which contain monomeric units that are covalently bonded to form larger molecules. The prefix 'bio' means they are biodegradable materials produced by living organisms. A wide variety of materials usually derived from biological sources such as microorganisms, plants, or trees can be described using the term "biopolymer". Materials produced by synthetic chemistry from biological sources such as vegetable oils, sugars, fats, resins, proteins, amino acids, and so on can also be described as biopolymer. As compared to synthetic polymers which have a simpler and more random structure, biopolymers are complex molecular assemblies that adopt precise and defined 3D shapes and structures. This is one important property which makes biopolymers active molecules in vivo. Their defined shape and structure are indeed keys to their function. For example, hemoglobin would not be able to carry oxygen in the blood if it was not folded in a quaternary structure.

The main property that distinguishes biopolymers from fossil-fuel-derived polymers is their sustainability, especially when combined with biodegradability. Biodegradable biopolymers from renewable resources have been synthesized to provide alternatives to fossil-fuel-based polymers. They are often synthesized from starch, sugar, natural fibers, or other organic biodegradable components in varying compositions. The biopolymers are degraded by exposure to bacteria in soil, compost, or marine sediment. Furthermore, subjecting biodegradable biopolymers to waste disposal by utilizing their characteristic of being degradable by the bacteria in the ground significantly reduces emission of CO_2 compared with conventional incineration. Therefore, attention is drawn to the use of biodegradable biopolymers from the viewpoint of global warming prevention. In recent years, with the critical situation of the global environment worsening due to global warming, the construction of systems with sustainable use of materials has been accelerated from the viewpoint of effectively using limited carbon resources and conserving limited energy resources. Furthermore, the cost of petroleum feedstocks has risen dramatically and there is a rising consumer interest in using "green" (or renewable resources) as the basis for consumer products.

One of the fastest-growing materials sectors in the past several years has been the production of polymers from renewable resources. Their development is fueled by the potential these polymers hold to replace fossil-fuel-based polymers. The main reasons for this drive can be summarized as follows:

- Inadequate fossil-fuel resources.
- Pricing instability of fossil fuel.
- Contribution of fossil fuel as a feedstock to climate change.
- Its occasional role as a political weapon.

- Its association with the waste disposal problem created by the fossil-fuel-derived polymers.

The main objectives of this chapter are to give a basic introduction of biopolymers, its classification and different sources, and method of preparation. This chapter also describes various physical and morphological characterization techniques used for biopolymers. Some widely used biopolymers are discussed in detail followed by applications of biopolymers in various fields especially in the field of nanotechnology.

Different Classes of Biopolymers

Biopolymers are classified into different ways based on different scales. Based on their degradability, biopolymers can be divided into two broad groups, namely biodegradable and non-biodegradable, and alternatively, into bio-based and non-bio-based biopolymers. On the basis of their polymer backbone, biopolymers can be classified roughly into the following groups: polyesters, polysaccharides, polycarbonates, polyamides, and vinyl polymers. These groups are again classified into several subgroups based on their origin.

Biopolymers can be classified, depending on the nature of the repeating unit they are made of, into three groups: (i) polysaccharides are made of sugars (e.g. cellulose found in plants), (ii) proteins are made of amino acids (e.g. myoglobin found in muscle tissues), and (iii) nucleic acids are made of nucleotides (DNA, genetic material of a given organism). Based on application, biopolymers can be classified as bioplastics, biosurfactant, biodetergent, bioadhesive, biofloculant, and so on.

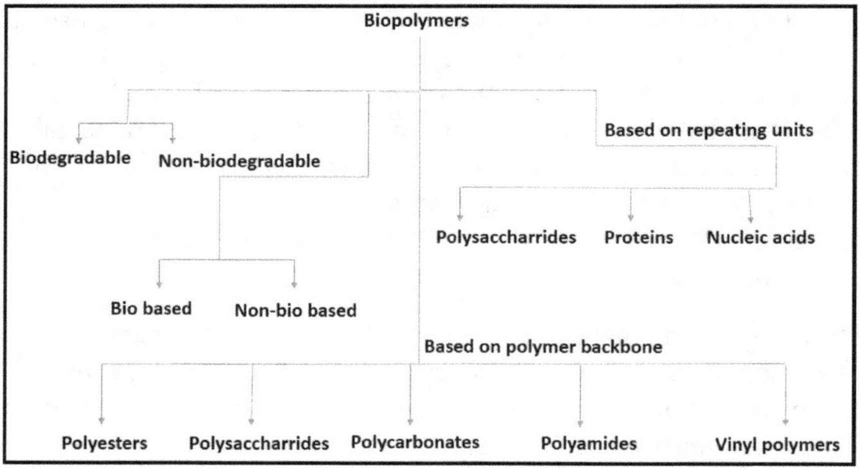

Figure: Different class of biopolymers.

Sources and Preparation of Biopolymers

Polymeric biomaterials are synthetically derived or modified polymers designed for various applications. Engineering the production of novel biopolymers in plants provides a truly bio renewable avenue for their synthesis. Like all polymer industries, these polymers are also produced in bulk and then shaped for a specific end use. Microorganisms also play an important role in producing a huge variety of biopolymers, such as polysaccharides, polyesters, and polyamides which range from viscous solutions to plastics. Their physical properties are dependent on the composition and molecular weight of the polymer. The properties of various biopolymers synthesized by the aid of microorganisms can be tailored by the genetic manipulation of microorganisms, thereby making it suitable for high-value medical application such as tissue engineering and drug delivery.

Biopolymers which are produced with the help of microorganisms require specific nutrients and controlled environmental conditions. They are produced either directly via fermentation or by chemical polymerization of monomers, which are in turn produced through fermentation. Most biopolymers are biocompatible with no adverse effects on biological systems. The mechanism of synthesis of biopolymers from bacterial origin is believed to be either as a result of their defense mechanism or as storage material. They can be degraded by natural processes, microorganisms, and enzymes so that it can be finally reabsorbed in the environment. Biopolymers or organic plastics are a form of plastics derived from renewable biomass sources such as vegetable oil, corn starch, pea starch, and so on. By shifting the focus more into the biopolymers, the conservation of fossil resources and reduction in CO_2 emissions can be achieved thereby promoting sustainable development. Among the microorganisms, algae serve as an excellent feedstock for plastic production because of their high yield and the ability to grow in a range of environments. The use of algae opens up the possibility of utilizing carbon and neutralizing greenhouse gas emissions from factories or power plants. Algae-based plastics have been a recent trend in the era of bioplastics compared to traditional methods of utilizing feedstock of corn and potatoes as plastics. While algae-based plastics are in their infancy, once they are into commercialization they are likely to find applications in a wide range of industries. Currently, microbial plastics are considered as an important source of polymeric material that has a great potential for commercialization. They can modify the flow characteristics of fluids, stabilize suspensions, flocculate particles, encapsulate materials, and produce emulsions.

Table: Some biopolymers, their sources, and preparation.

Polymer	Method/ Source	Example of bacteria used for synthesis.
Hybrid plastics	Adding denatured algae biomass to petroleum-based plastics like polyurethane and polyethylene as fillers	Filamentous green algae, Cladophorales.
Cellulose-based plastics	Biopolymer of glucose	30 % of the biomass produced after extraction of algal oil is known to contain cellulose.
Poly-lactic acid (PLA)	Polymerization of lactic acid lactic acid	Bacterial fermentation of algal biomass.
Bio-polyethylene	Ethylene produced from ethanol, by a chemical reaction called cracking. Ethanol derived from natural gas or petroleum	Bacterial fermentation of algal biomass.
Poly esters	Biomass	Bacteria like Akaligenes eurrophus, E. coli, etc.

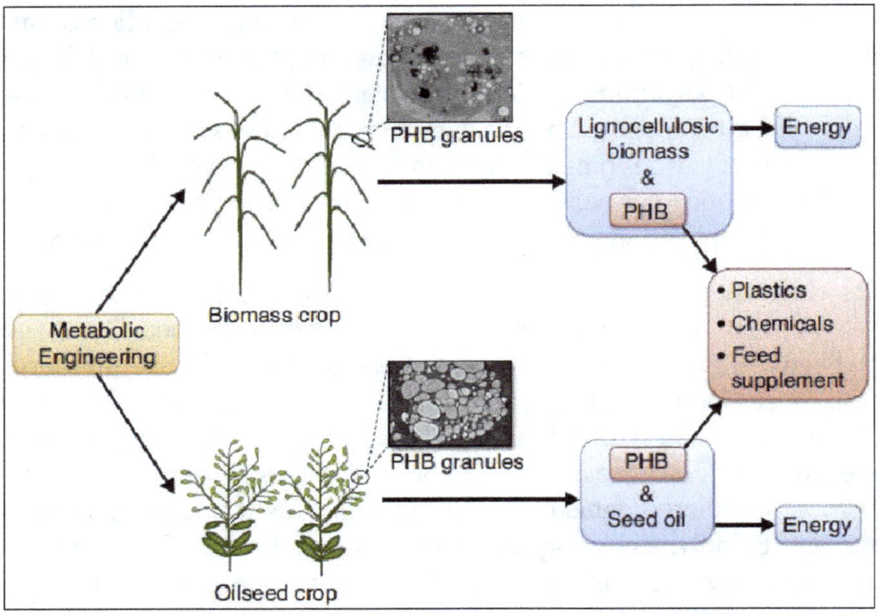

Figure: Schematic representation of the production of polyhydroxybutyrate (PHB) in both biomass and oilseed crops.

Structure of Biopolymers

The structure of biopolymers has been a subject of investigation ever since they were discovered. So far, the most accurate methods for structure determination have been X-ray diffraction techniques. These techniques are, however, restricted to crystallizable molecules, and it has not been established if the molecular structure in a crystal is identical to that in solution. Many methods have been developed or adopted to probe biopolymer properties in solution. The composition and sequence of the monomer units of biopolymers determine their so-called primary structure. Conformation, that is, one or another spatial form of the biopolymer molecules, is determined by their primary structure. Depending on the chemical structure and external conditions, the molecules may be in one or more preferred conformations.

Specific form and structure provide functionality to biological elements microscopically and macroscopically. The combination of cell and cell tissues results in the outward appearance of biological organisms. The causes for the shape of these building blocks of life are manifold. Biopolymers are important for the organization and functionality of cells. Prominent examples are the DNA as the carrier of the genetic code and the cytoskeletal filaments building the scaffold of the cell. The unique property of DNA depends on its equilibrium polymer conformations based on its internal characteristics like flexibility, topology and filament diameter and external constraints such as confinement. Any variation in these properties will alter the genetic characteristics of that living system which shows the significance of structure of biopolymers in living systems.

Different Characterization Techniques

Most of the techniques used for the characterization of polymers can be utilized for the characterization of biopolymers. Characterization of biopolymers has two purposes:

- Development of parameters for processing.
- Determination of end-use performance characteristics.

Through characterization, the most important properties of interest are molecular mass, polydispersity, size, degree of association, conformation, interaction, and so on. Some of the characterization techniques which are used for the analysis of biopolymers are listed.

Fourier Transform Infrared Spectroscopy (FTIR)

FTIR can be used to study the various functional groups and interactions present in biopolymers. Interactions like hydrogen bonding, amide linkage, etc., can be easily detected by analyzing the spectra. Among various techniques, attenuated total reflection (ATR) sampling device-based Fourier transform infrared (ATR-FTIR) spectroscopy is considered as a potential one. It requires minimal sample preparation, permits routine analysis at both laboratory and on-field environments, and is easy to operate. Visible and near-infrared (VIS–NIR) spectroscopy is a well-established alternative for measuring constituents of biological materials. The NIR spectrum is influenced by the different vibrational modes of the molecules, which are caused by their interaction with electromagnetic radiation absorbed at specific wavelengths. This technique can be used for the identification of several complex components such as proteins and carbohydrates in biological materials. Chemometrics allows the extraction of relevant information contained in the spectra for the development of calibration models that allow the classification and prediction of organic sample's composition. The thermotropic phase behavior of a suite of newly developed self-forming synthetic biopolymers has been investigated by variable-temperature Fourier transform infrared (FT–IR) absorption spectroscopy.

Various studies have extensively used the FTIR technique to explore the physical and mechanical properties of biopolymers. Two-dimensional (2D) correlation analysis was applied to characterize the ATR spectral intensity fluctuations of immature and mature cotton fibers. FTIR analysis can also be used to verify the interactions between various biopolymers in its composites.

Figure shows FTIR spectra of chitosan-starch films, keratin, and the composites synthesized with 20 %. Figure below shows the band at 3400 cm^{-1} which is assigned to the hydrogen-bonded hydroxyl groups due to the presence of starch in chitosan-starch film. It shows that the main vibrations attributed to the keratin structure were identified in the region around 3300 cm^{-1} and correspond to a range of amide bands. Furthermore, the peak at 2945 cm^{-1} is assigned to the asymmetric vibration of the CH of the methyl group. The main peaks related to chitosan-starch films are seen at 700–770 cm^{-1} which correspond to the saccharide structure, and the three-characteristic peaks between 995 and 1150 cm^{-1} are attributed to C–O bond stretching.

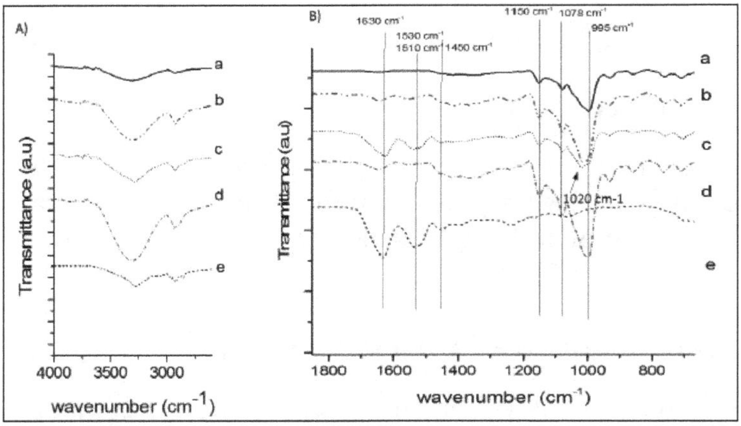

Figure: IR spectra of the composite films: A) from 4000–2500 cm-1; B) from 1800–700 cm-1. (a) chitosan-starch film; (b) chitosan-starch-short biofiber 20%; (c) chitosan-starch-long biofiber 20 %; (d) chitosan-starch-ground quill 20 %; (e) keratin.

In Figure above A(e) the band at 1630 cm^{-1} is assigned to the C=O group and the peaks at 1530 cm^{-1} and 1510 cm^{-1} are attributed to the in plane bending of NH group, while the signal at 1,430 cm^{-1} is related to bending of the CH_3 group for keratin. FTIR results correlate to the thermomechanical properties and the morphological features of the composite. The spectrum indicates that a better storage modulus is obtained for composites with short and large fibers than the composite with ground quill. Also better distribution is obtained in these composites.

X-ray Diffraction Analysis

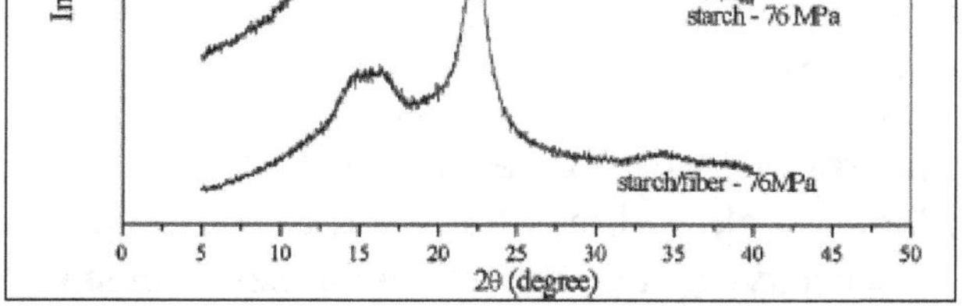

Figure: X-ray diffraction profile for native starch, starch (76 MPa) and starch/fiber (76 MPa).

X-ray diffraction techniques can be successfully used to study the crystallinity of the biopolymers. Figure represents the X-ray diffraction studies of native wheat starch, compressed starch foam, and compressed starch/fiber foam. The native wheat starch had a semi-crystalline structure with peaks typical of A-type crystals. The gelatinization/compression processes during the composite preparation cause decrease of crystallinity resulting in diffractograms exhibiting a broad amorphous peak with no significant residual A-type crystallinity. The starch granule structure was disrupted by the processes of gelatinization and high compression which is clearly visible from the spectrum. Small peaks were found around 2θ 12.0, 17.3, and 19.5, which is characteristic for crystallographic parameters of VH-type crystals for destructured starch prepared from solution. For the starch/fiber sample, the spectrum obtained indicates that the crystallinity of the fiber overlaps with the gelatinized starch.

NMR Spectroscopy

The molecular dynamics of water adsorbed on biopolymer surfaces and functional groups of biopolymers can easily be determined by using the NMR technique. NMR spectroscopic technique relaxation times and their dependence on temperature and magnetic field have proved to be a useful source of information for molecular structure, phase change, conformational exchange, solubility, and diffusion of biopolymers. Many studies of spin–lattice relaxation time (T_1), spin–spin relaxation time (T_2), and spin–lattice relaxation time in the rotating frame (T_{1q}) of different biopolymers have been already reported. The molecular origin of nuclear spin relaxation has been examined in cellulose and its derivatives. In order to understand the behavior of main functional groups, the effect of temperature on nuclear-spin relaxation has been probed extensively in biopolymers in cell walls. The temperature dependence of relaxation times of starch chain has been studied in order to investigate structure–hydration relationships in the biopolymer.

Thermal Analysis

Thermal analysis (TA) method has become an indispensable analytical technique for the characterization of polymers. The characterization technique provides relevant information about thermal and chemical stability, phase transition temperatures and kinetics, rheology and molecular relaxation times of the polymers. It provides information regarding the relation between the polymers chemical structure and their physical properties. The properties of processed polymer depend on morphology and molecular chain orientation of the polymer backbone. In conventional thermal analysis techniques, the method usually involves the monitoring of material properties as a function of temperature. Differential scanning calorimetry (DSC) is one important method used to study the thermal behavior of biopolymers.

Figure below shows the standard/conventional DSC curves of the neat blend and blend/clay composites. All of the samples showed a glass transition with enthalpy relaxation at approximately 55°C. There was some apparent cold crystallization for the samples that could not be clearly separated from the melting endotherm. The melting endotherms appeared at approximately 150°C, and the values could not be accurately determined because of the apparent influence of the cold crystallization exotherms. It was clear that the area under the cold crystallization and melting peaks (enthalpy of cold crystallization and melting, respectively) was higher for the control sample and the samples with less to moderate hydrophobic clays when compared with that of the composites with hydrophobic to extremely hydrophobic clays. This indicated the tendency of the more hydrophobic clays to hinder the re-arrangement of the PLA molecules into organized crystalline structures. Such a hindrance was highest in the case of the PLA-ST-C93A composite because of the high level of compatibility between two phases.

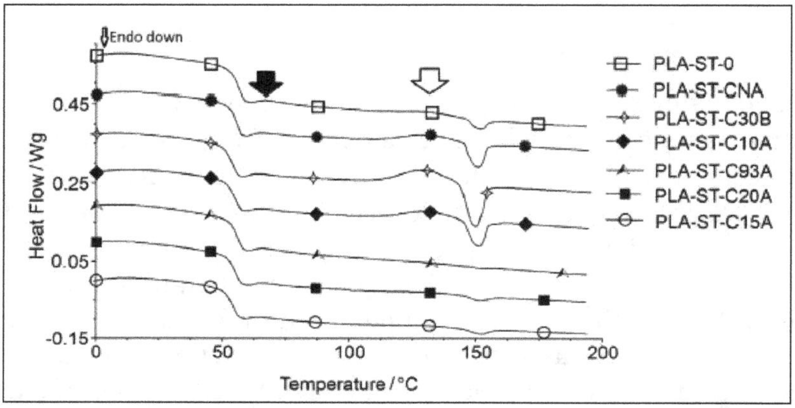

Figure: DSC thermograms of PLA/butylated starch blend and blend/MMT composites with MMTclays of different hydrophobicity. The samples had a clay and starch content of 4.5 wt % and 25.5 wt % of the total composite weight, respectively. The control sample PLA-ST-0 had no clay.

VIS–NIR Spectroscopy

There is an increasing interest in the use of polysaccharides and proteins for the production of biodegradable films. VIS–NIR spectroscopy is a reliable analytical tool for objective analyses of biological sample attributes. It can be used to investigate the compositional characterization of biodegradable materials and correlation of their mechanical properties. Partial least square regression can be used to investigate the correlation between spectral information and mechanical properties. Large-scale production of films requires fast and non-invasive analytical methods in order to assess film composition and mechanical properties. VIS–NIR spectroscopy is a well-established alternative as a non-destructive process analytical technology for measuring constituents of biological materials. The NIR spectrum is influenced by the different vibrational modes of the molecules, which are caused by their interaction with electromagnetic radiation absorbed at specific wavelengths. This technique has been used for the identification of several complex components such as proteins and carbohydrates in biological materials. Chemometrics allows the extraction of relevant information contained in the spectra for the development of calibration models that allow the classification and prediction of the composition of organic samples.

Scanning Kelvin Probe Microscopy (SKPM)

SKPM has become a very strong tool in material science, giving up a surface mapping of electrostatic potential. Electrical properties have an important role in the rejection of charged species in membrane applications such as water softening and heavy metal removal. Increasing membrane charge may improve ion rejection for Nano filtration (NF) and reverse osmosis (RO) processes. If membrane is electrically charged, electrostatic interaction becomes significant for ion removal. Nanocomposites which presented the best flows were characterized by SKPM for getting information about their electrical potential.

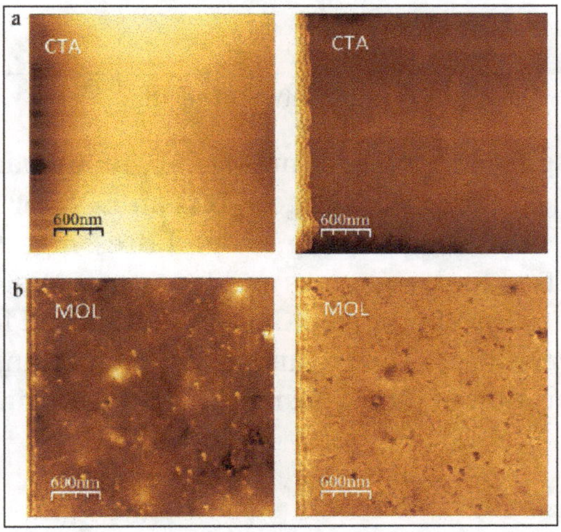

Figure: SKPM views of CTA (cellulose triacetate) and Nano filtration composites MOL (Lignin). Topography and surface potential mapping are shown at the left and at the right side of the figure, respectively.

Figure above shows SKPM image of cellulose triacetate and lignin. The dark zones in the images correspond to a potential difference between surface and AFM tip, whereas no difference corresponds to bright zones. For neat CTA (cellulose triacetate) membranes, no indication of electrical potential at the surface of the material was found. When lignin is incorporated into the polymer membrane, the resulting Nano filtration membranes showed a surface electrical potential.

Frequently Studied Biopolymers

Gelatin

Gelatin is a protein made from animal products. It is produced by the partial hydrolysis of collagen obtained from skin, bones, and tissues of animals. They are mostly derived from type I collagen containing no cysteine. There are mainly two types of gelatin based on their synthesis procedure and composition. Type A, which is obtained by acid hydrolysis of collagen, has 18.5 % of nitrogen and type B obtained by the alkaline hydrolysis of collagen has 18 % of nitrogen as they do not contain amide groups. At elevated temperatures, it melts and exists as coils and on lowering the temperature, it undergoes coil to helix transformation. Collagen is resistant to proteases, whereas gelatin is susceptible to these enzymes. Due to the presence of many functional groups like $-NH_2$, –SH, and –COOH, gelatin can be modified with biomolecules and nanoparticles for various applications.

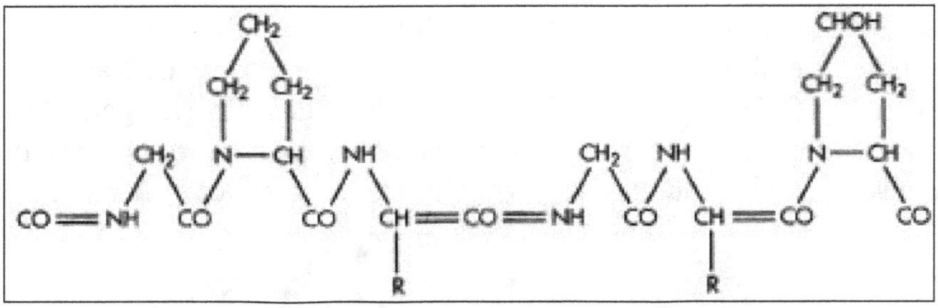

Figure: Structure of gelatin.

It is hemostatic, pro-angiogenic, non-immunogenic, biodegradable, and biocompatible and can be cross-linked to form hydrogels. It is an ECM protein with wide number of applications in bone tissue engineering, wound dressing, drug delivery, and gene transfection. Gelatin is used for weight loss and for treating osteoarthritis, rheumatoid arthritis, and brittle bones (osteoporosis). Gelatin is also used for improving hair quality and to shorten recovery after exercise and sports-related injury. In manufacturing, gelatin is used for preparation of foods, cosmetics, and medicines.

Starch

Starch is a semi-crystalline polymer composed of amylose and amylopectin molecules. The primary carbohydrate used for energy storage in plants is starch, the equivalent of glycogen in animals. Both starch and glycogen are polymers of glucose, with different glycosidic linkages between glucose monomers and different degrees of branching. Amylose is a linear polymer composed of (1–4)-linked a-D-glucopyranosyl units while amylopectin is a very large molecule made up of (1–4)-linked a-D-glucopyranose chains with a-(1–6) branches. Starch is one of the most abundant natural polymers and it is cheap and degradable. Properly processed starch can be made into various commercial products. Due to their sensitivity to moisture and poor mechanical properties, they have only found use in various niche markets. Studies have shown that by incorporating fibers, the mechanical and barrier properties of starch products can be improved. Nano and micro structured fibers when dispersed in the polymer matrix while preparing the composite, will improve its tensile strength and flexibility.

The significance of starch in plants and humans are undeniable. While plants consider it as a storage of energy, humans need it as a vital part of their diet and also for various commercial products. Starch and its derivatives have already been explored in various applications like in the manufacture of paper, textiles and adhesives. Their biodegradable and renewable nature make them an environmental friendly alternative to the use of synthetic additives in many other products, including plastics, detergents, pharmaceutical tablets, pesticides, cosmetics, and even oil-drilling fluids.

Cellulose

The monomer unit in cellulose is β-1, 4 linked glucan chains with hydrogen bonds formed between hydroxyl groups and oxygen atoms both within a single glucose chain and between neighboring chains. Hydrogen bonding and van der Waals forces aggregate glucan chains together side by side to form cellulose microfibrils. These cellulose microfibrils are then stacked together to form crystalline cellulose. The polar –OH groups present in the cellulose chain form many hydrogen bonds with OH groups on adjacent chains, bundling the chains together. The stacking of chains is so regular that it forms hard, stable crystalline regions, and this gives the bundled chains more stability and strength. The length of the chain varies greatly from a few hundred sugar units in wood pulp to over 6000 for cotton. It is the most common natural organic polymer which is considered as an almost inexhaustible source of raw material for the increasing demand for environment friendly and biocompatible products. Due to the strong hydrogen bonding, it is insoluble in common solvents. As a result, it is chemically modified into ether, ester, and acetal derivatives.

Cellulose, the most abundant biological material on Earth, is also composed of glucose monomers but joined by beta glycosidic bonds, giving it a straighter shape that packs closely and provides mechanical strength in wood.

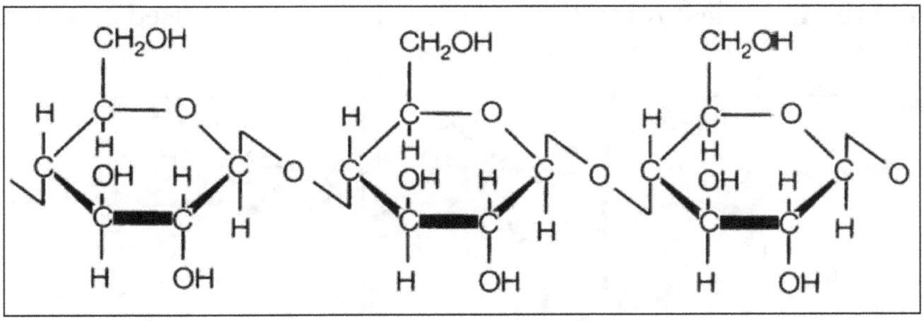

Figure: The structural representation of cellulose.

The remarkable strength of wood is due to cellulose which is a long chain of linked sugar molecules. Plant cell walls, which are the basic building block for textiles and paper, also have cellulose as main component. The purest natural form of cellulose is cotton which is used in many textile applications.

Recently, cellulose has been excessively used in the form of nanofibrils. Cellulose nanofibrils (CNF) or nanocellulose is typically generated by mechanical grinding or high-pressure fluidization of cellulose to remove its lignin content. CNF consists of very thin (5–20 nm) and long (several μm) fibrils with high aspect ratio. At low concentrations, it forms a transparent gel-like material which can be used for producing biodegradable and environmentally safe, homogeneous, and dense films for various applications, especially in biomedical field. Extraction of CNF has been reported from various sources like coir, banana, sugar beet, hemp, softwood, and hardwood pulps. After using various plasticizers, various thermal, mechanical, barrier, and physical properties of the cellulose can be improved so that it can be used in various commercial applications. Plasticizing improves properties like grease proofness, and high barrier against oxygen transmission especially at dry conditions. All cellulose-synthesizing organisms including bacteria, algae, tunicates, and higher plants have cellulose synthase proteins, which catalyze the polymerization of glucan chains.

Chitin and Chitosan

Chitin is a nitrogen-containing polysaccharide, related chemically to cellulose that forms a semitransparent horny substance, and is a principal constituent of the exoskeleton, or outer covering, of insects, crustaceans, and arachnids. The structure of chitin is comparable to the polysaccharide cellulose, which consists of β-(1 → 4) linked nacetylglucosamine, forming crystalline nanofibrils or whiskers. Chitin is the second most abundant naturally occurring polysaccharide found in the outer shell of crustaceans and insect exoskeleton. Chitin is insoluble in most of the organic and inorganic solvents. Chitosan, the de-acetylated derivative of chitin, is a cationic polysaccharide that can form polyelectrolyte complexes with other polysaccharides and is the second most abundant natural biopolymer after cellulose. It consists of β-(1 → 4)-linked D-glucosamine and N-acetyl-D-glucosamine. Due to easy applicability, water sorptivity, oxygen permeability, blood coagulating property, and cytokine induction (interleukin-8), which activate the fibroblast migration and proliferation, it is used as wound dressing material.

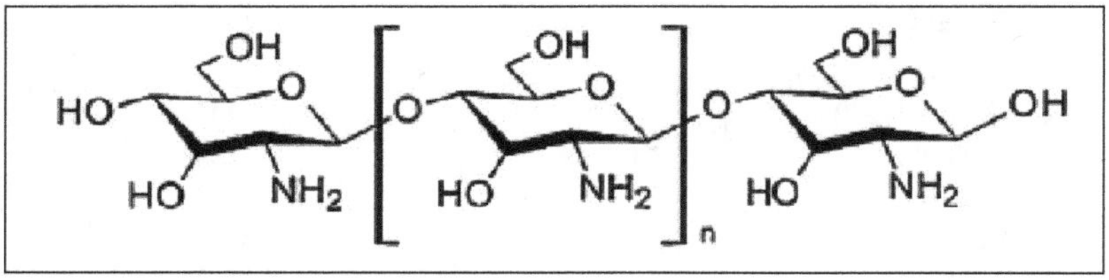

Figure: The structure of (a) chitosan biopolymer.

Chitosan, due to its good film forming property, has been widely used for casting free-standing films. It can be applied as a coating on various surfaces to improve the permeability of the surface against moisture and various gases. The water vapor permeability of the paper increased as a result of the chitosan coating. Chitosan coatings can be made on to NH_3- or CO_2-plasma activated polypropylene films, thereby making the film surfaces hydrophilic and prevent oxygen, carbon dioxide, and ethylene transmission. Barrier properties can be further improved by using nanoclays incorporated into chitosan.

Polylactic Acid (PLA)

Polylactic acid or polylactide (PLA, Poly) is a biodegradable thermoplastic aliphatic polyester derived from renewable resources, such as corn starch, tapioca roots, chips, starch, or sugarcane. PLA is obtainable primarily by the ionic polymerization of lactide, a ring closure of two lactic acid molecules. At temperatures between 140 and 180 °C and under the action of catalytic tin compounds (such as tin oxide), a ring-opening polymerization takes place. Lactide itself can be made through lactic acid fermentation from renewable resources such as starch by means of various bacteria. PLA can also be produced directly from lactic acid by polycondensation. However, this process yields low molecular weight polymers, and the disposal of the solvent is a problem in the industrial production. The properties of PLA primarily depend on the molecular mass, the degree of crystallinity, and possibly the proportion of co-monomers. A higher molecular mass raises T_g, as well as T_m, tensile strength, elastic modulus, and lowers the strain after fracture. Due to the CH_3 side group, the material has water-repellent or hydrophobic behavior. PLA is soluble in many organic solvents, such as dichloromethane or the like. PLA has higher transparency than other biodegradable polymers, and is superior in weather resistance and workability.

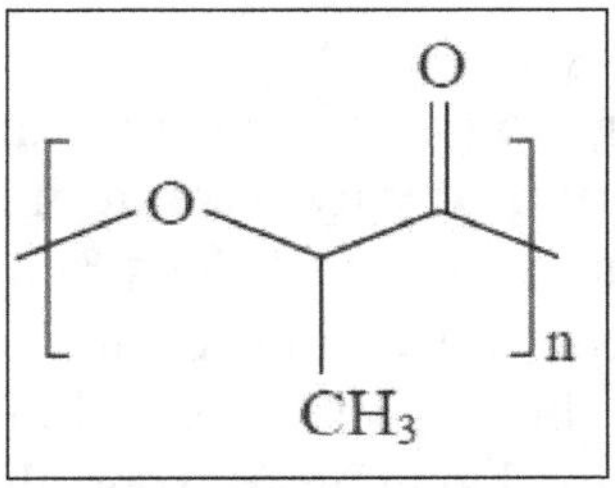

Figure: Structural formula of PLA.

Poly(e-Caprolactone) (PCL)

Poly(e-caprolactone) (PCL) is a biodegradable polyester which is commonly used for preparing scaffolds for various tissue engineering applications. It has a low melting point of around 60 °C and a glass transition temperature of about −60 °C. It is used in the manufacture of specialty polyurethanes with good water, oil, solvent, and chlorine resistance. PCL can be used as an additive for resins to improve its impact resistance and their processing characteristics. PCL is compatible with a range of other materials which enable it to be mixed with other biodegradable polymers to lower its cost and increase biodegradability. It can also be used as a polymeric plasticizer to other polymers like PVC. Polycaprolactone is also used for splinting, modeling, and as a feedstock for prototyping systems such as a 3D printer. PCL is biocompatible, biodegradable polymer which has shown to be capable of supporting a wide variety of cell types. PCL is also a well-known FDA-approved biomaterial which is most widely used in biomedical field. The PCL is intrinsic hydrophobic in nature. Its poor surface wetting and interaction with the biological fluids avoid cells adhesion and proliferation. PCL is prepared by ring opening polymerization of ε-caprolactone in the presence of a catalyst. The significance of PCL in various tissue engineering applications increased because it is degraded by hydrolysis of its ester linkages in physiological conditions. This property makes it useful for various implantable biomaterials. Since its degradation rate is lower than that of polylactide, it is proposed to be used as long-term implantable devices.

Poly(Vinyl Alcohol) (PVA)

Poly(vinyl alcohol) (PVOH or PVA) is a white, colorless and odorless, water-soluble synthetic polymer with molecular formula $[CH_2 CH(OH)]_n$. It is commercially available as beads or as solutions in water. It is used as thickener in glues, paper-making, sizing agent in textiles, water-soluble films useful for packing, and a variety of coatings. Polyvinyl alcohol is used as an emulsion polymerization aid, as protective colloid, to make polyvinyl acetate dispersions. PVA is prepared by first polymerizing vinyl acetate, and the resulting polyvinyl acetate is converted to the PVA. PVA has a wide range of potential applications in optical, pharmaceutical, medical, and membrane fields. It is a water-soluble polymer and allows the development of environment-friendly material processes.

Polyvinyl Acetate (PVAc)

Polyvinyl acetate (PVAc) belongs to polyvinyl ester family and is a thermoplastic resin produced by the polymerization of vinyl acetate monomer. Its general formula is $[CH_3COOCHCH_2]_n$ with a solids content of 50–55 % in water. Most polyvinyl acetate emulsions available in market contain various co-monomers such as n-butyl acrylate, 2-ethyl hexyl acrylate, ethylene, dibutyl maleate, and dibutyl fumarate. Polymerization of vinyl acetate with ethylene can also be used to produce solid vinyl acetate/ethylene copolymers with more than 50 % vinyl acetate content. Polyvinyl alcohol (PVOH) is produced by methanolysis or hydrolysis of polyvinyl acetates. The reaction can be controlled to produce any degree of replacement of acetate groups. Polyvinyl acetate has immense applications in various fields like biomedical, synthesis of metal nanoparticles, sensing activity, and so on.

Collagen

Collagen is the most abundant protein in human. It is found in the bones, muscles, skin, and tendons, where it forms a scaffold to provide strength and structure. Endogenously produced, collagen plays numerous important roles in health, with the breakdown and depletion of the body's natural collagen associated with a number of health problems. Exogenous collagen is supplied to the body for medical and cosmetic purposes, including helping with healing and repairing of the body's tissues. Around 30 % of the protein in the human body is collagen and 80–90 % of the collagen in the body consists of types I, II, and III. Among this, type I collagen fibrils are stronger than steel hence is the substance that holds the whole body together. Collagen gives the skin its strength and structure, and also plays a role in the replacement of dead skin cells. Collagen in medical products can be derived from human, bovine, porcine, and ovine sources. Collagen dressings attract new skin cells to wound sites. Controllable factors that damage the production of collagen include sunlight, smoking, and high sugar consumption.

Poly(3-Hydroxybutyrate-co-3-Hydroxyvalerate) PHBV

Poly(3-hydroxybutyrate-co-3-hydroxyvalerate) PHBV, is a polyhydroxyl- alkanoate-type thermoplastic linear aliphatic polyester. PHBV is synthesized by bacteria as an intercellular carbon and energy storage compounds under growth limiting conditions. It is biodegradable, nontoxic, and biocompatible plastic which can be produced from glucose and propionate by the recombinant Escherichia coli strains. PHBV is a copolymer of 3-hydroxybutanoic acid and 3-hydroxypentanoic acid and can also be synthesized from butyrolacone and valerolactone in presence of oligomeric aluminoxane as catalyst. PHBV has many applications especially for the development of implanted medical devices for dental, orthopedic, hernioplastic, and skin surgery. Various potential medical devices like bioresorbable surgical sutures, biodegradable screws and plates for cartilage and bone fixation, biodegradable membranes for periodontal treatment, surgical meshes with PHBV coating for hernioplastic surgery, wound coverings, and the like have been developed using this biopolymer.

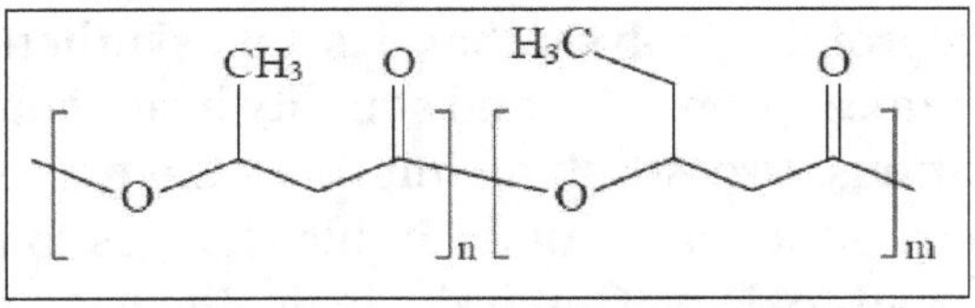

Figure: Structural formula of PHBV.

Applications of Biopolymers

Biopolymers, due to its biocompatible and biodegradable nature, can be used to improve the performances of other biologically active molecules in a product. They can also be modified to suite various potential applications which include the following:

Synthesis of Nanomaterials

Nanotechnology is the science of nanomaterials which deals with its synthesis, characterization, and applications. Researchers are currently focusing on developing more eco-friendly processes for the synthesis of nanoparticles. The main focus for the synthesis protocol has shifted from physical and chemical processes towards "green" chemistry and bioprocesses. Metal nanoparticles, due to their quantum size effects, possess various novel properties. However, most of their synthesis protocol imposes a major threat to the environment. In common synthetic methods, the reducing agents used which include organic solvents and toxic-reducing agents like hydrazine, N-dimethylformamide, and sodium borohydride are considered to be highly toxic for the environment. All these chemicals are highly reactive and pose potential environmental and biological risks. With the increasing interest in minimization/elimination of waste and adoption of sustainable processes, the development of green chemistry approaches is desirable. Biopolymers have been extensively used as capping and reducing agent for the synthesis of various nanoparticles. Biopolymers like chitosan, heparin, soluble starch, cellulose, gelatin, PVA, PVP, and so on can be used to replace various toxic regents in synthesizing different nanoparticles.

Biomedical Applications

In recent years, biopolymer materials have aroused great interest because of their biomedical applications, such as those in tissue engineering, pharmaceutical carriers, and medical devices. A common biopolymer, gelatin, was widely applied in medicine for dressing wounds, as an adhesive, and so on. Porous gelatin scaffolds and films were produced with the help of solvents or gases as simple porogens, which enable the scaffolds to hold drug or nutrients to be supplied to the wound for healing. Electrospun PLGA-based scaffolds have been applied extensively in biomedical engineering, such as tissue engineering and drug-delivery system. MWCNT-incorporated electrospun nanofibers with high surface area-to-volume ratio and porous characteristics have also shown potential applications in many aspects of tissue engineering.

Biomaterials made from proteins, polysaccharides, and synthetic biopolymers are preferred but lack the mechanical properties and stability in aqueous environments necessary for medical applications. Cross-linking improves the properties of the biomaterials, but most cross-linkers either cause undesirable changes to the functionality of the biopolymers or result in cytotoxicity. Glutaraldehyde, the most widely used cross-linking agent, is difficult to handle and contradictory views have been presented on the cytotoxicity of glutaraldehyde-cross-linked materials.

Food Industry

Replacing the oil-based packaging materials with bio based films and containers might give not only a competitive advantage due to more sustainable and greener image but also some improved technical properties. Biopolymers are currently used in food coatings, food packaging materials, and encapsulation matrices for functional foods. They provide unique solutions to enhance product shelf life while also reducing the overall carbon footprint related to food packaging. Within food-related applications, these biobased materials are particularly useful in three main areas: food packaging, food coating, and edible films for food and encapsulation. The most commercially viable materials in food packaging are certain biodegradable polyesters and thermoplastics like starch, PLA, PHA, and so on, which can be processed by conventional equipment. These materials are already used in a number of monolayer and multilayer applications in the food-packaging field. Starch and PLA biopolymers are potentially the most attractive types of biodegradable material. This is due to the balance of their properties and the fact that they have become commercially available. PLA is of particular interest in food packaging, due to its excellent transparency and relatively good water resistance. The challenge for these specific biomaterials is to improve their barrier and thermal properties so that they perform like polyethylene terephthalate (PET). Other materials extracted from biomass resources, such as proteins (e.g., zein), polysaccharides (e.g., chitosan), and lipids (e.g., waxes), also have excellent potential as gas and aroma barriers. The inherent high rigidity and the difficulty of processing them in conventional equipment are the main drawbacks of these types of materials. The hydrophilic nature of most of the biopolymers affects their use as high-end products. The absorption of moisture causes plasticization of these materials thereby deteriorating the barrier properties of these materials.

Renewable polymers have also been used for encapsulation purposes. Encapsulation has previously been described as a technology to protect sensitive substances against the influences of adverse environments. The term "microencapsulation" refers to a defined method of wrapping solids, liquids, or gases in small capsules, which can release their contents under specific circumstances. Such technologies are of significant interest to the pharmaceutical sector. The increasing interest in edible films and coatings using biopolymers is due to their ability to incorporate a variety of functional ingredients. Plasticizers, such as glycerol, acetylated monoglycerides and polyethylene glycol, which are used to modify the mechanical properties of the film or coating, make significant changes to the barrier properties of the film. However, the major advantage of coatings is that they can be used as a vehicle for incorporating natural or chemical active ingredients, such as antioxidants and antimicrobial agents, enzymes, or functional ingredients, like probiotics, minerals, and vitamins. These ingredients can be consumed with the food, thus enhancing safety, nutritional, and sensory attributes. Edible films can be used as flavor or aroma carriers in addition to providing a barrier to aroma loss.

Chitosan has shown great potential as an antimicrobial packaging agent to preserve food against a wide variety of microorganisms. Incorporating antimicrobial compounds into edible films or coatings provides a novel way to improve the safety and shelf life of ready-to-eat foods. Lysozyme is one of the most frequently used antimicrobial enzymes in packaging materials, since it is a naturally occurring enzyme. Biopolymers such as amylose, when mixed with plasticizers have excellent potential in forming thin films for various food and packaging applications. Starch has high sensitivity to relative humidity (RH) due to its hydrophilic nature and this can be reduced by introducing plasticizers which enhances the flexibility of the matrix. But this technique also has some limitations due to the complex interactions between the hydrophilic plasticizers and the starch. An "anti-plasticization" process takes place with increased stiffness of the matrix if the structure of the plasticizer molecule and the polymer matrix is not compatible.

Packaging Applications

Currently, the most commercially viable materials in food packaging are certain biodegradable polyesters, which can be processed by conventional equipment. These materials are already used in a number of monolayer and multilayer applications in the food-packaging field. Among the most widely researched thermoplastics, the sustainable biopolymers used in monolayer packaging include starch, PHA, and PLA. Starch and PLA biopolymers are potentially the most attractive types of biodegradable material. This is due to the balance of their properties and the fact that they have become commercially available. The challenge for these specific biomaterials is to improve their barrier and thermal properties so that they perform like polyethylene terephthalate (PET). Other materials extracted from biomass resources, such as proteins (e.g., zein), polysaccharides (e.g., chitosan), and lipids (e.g., waxes), also have excellent potential as gas and aroma barriers. The main drawbacks of these types of materials are their inherently high rigidity and the difficulty of processing them in conventional equipment.

For bio-based food-packaging applications, the most important parameter to be considered is its barrier properties. Hydrophilic polymers usually have poor moisture resistance which cause water vapor transmission through packaging and thus affect the quality of foods. This results in shorter shelf lives, increased costs, and eventually more waste. Another technique to improve the barrier properties of biopolymers is to add various nanofillers like nanoclays, metal oxide nanoparticles, and so on. Among the bioplastics, polyglycolic acid (PGA) has excellent barrier properties thus it's one of the most promising new commercially available barrier polymers. Its precursor, glycolic acid, can now be produced via a natural metabolic route, the glyoxylate cycle.

Water Purification

Safe drinking water is a significant, but simple indicator of development. Nanotechnology has shown promising developments in providing safe drinking water through effective purifying mechanisms. Several nanomaterials have already proved to have antibacterial and antifungal properties. Developing affordable materials which can constantly release these antibacterial materials like silver nanoparticles to water is an effective way of providing microbially safe drinking water for all. Developing various nanocomposites with functional materials which can scavenge various toxic metals like arsenic, lead, etc., from water together with the antibacterial agents can result in affordable water purifiers that can function without electricity. The main challenge in this technology is developing stable materials which can release nanoparticles continuously overcoming the scaling on nanomaterials caused by various complex species present inside water.

A relatively new biopolymer, chitosan, shows superior performance where many conventional polymers fail. It is a versatile polymer with applications in water treatment, biomedical and dietary supplement industries. Chitosan is used as a flocculant in water treatment processes and will biodegrade in the environment over periods of weeks or months rather than years. Compared to chitosan, many aggressive and cheap synthetic flocculants are available but they leave a residual impact on the environment. Chitosan removes metals from water by forming chelates.

Chelation is a process by which multiple binding sites along the polymer chain bind with the metal to form a metal cage like structure, to remove it from a solution. This property of chitosan together with its biodegradability make it an eminent candidate for treating difficult industrial storm water and waste water, where conventional methods failed to reduce the contaminant levels. Porous GO-biopolymer gels can efficiently remove cationic dyes and heavy metal ions from wastewater. Nanocrystalline metal oxyhydroxide-chitosan granular composite materials prepared at near room temperature through an aqueous route was also efficient in water purification. Nano fiber membranes can improve the water-filtration process without adversely affecting environment. Combining various nanomaterials together with biopolymers can effectively restrict the formation of biofilms on the polymer surface.

Classification of Polymers

Polymers are classified on different basis and types as explained below:

Classification based on the Source of Origin

- Natural Polymers are those which are found in animals and plants. This includes cellulose, starch, proteins, resins etc.

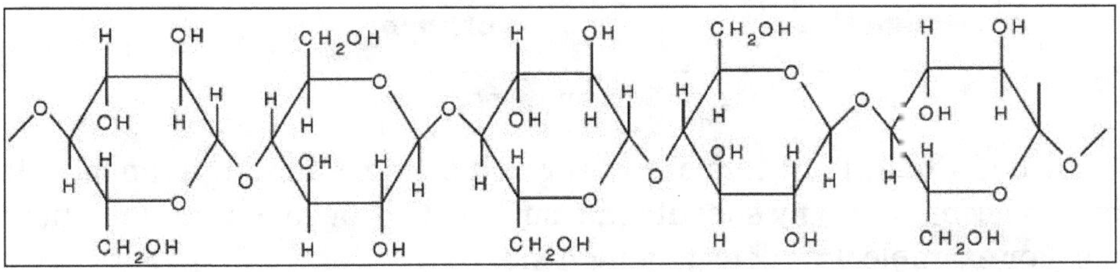

Figure: Structure of cellulose.

- Semi-synthetic Polymers are chemically modified natural polymers such as cellulose nitrate, cellulose derivatives such as rayon etc.

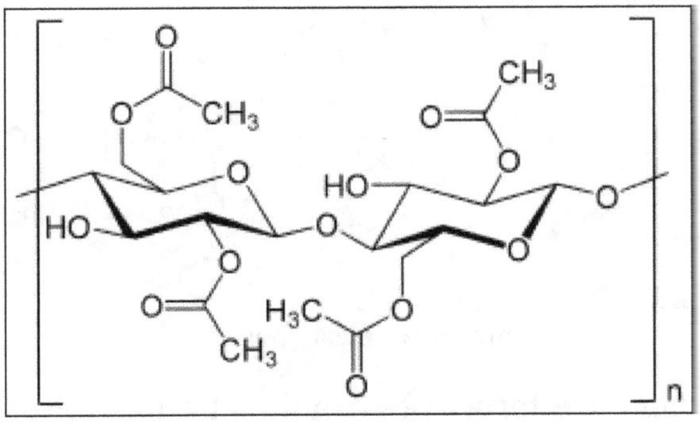

Figure: Structure of rayon.

- Synthetic Polymers are man-made polymers. It includes polythene, synthetic fibers, synthetic rayon etc. They are man-made products.

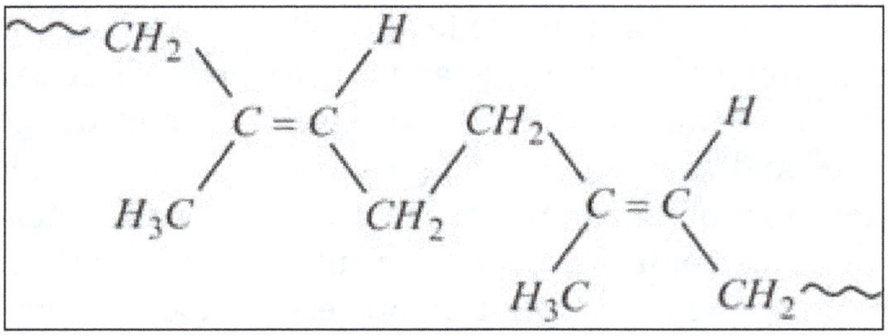

Figure: Structure of rubber.

Classification based on the Structure of Polymers

- Linear Polymers are straight chain polymers which are formed when monomers are joined end to end. For example, polythene, polyvinyl chloride etc.

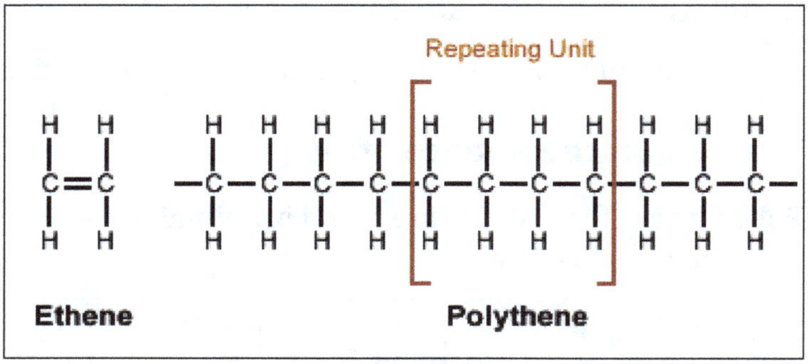

Figure: Structure of polythene.

- Branched Chain Polymers are linear polymers containing branches. It occurs by replacement of any substituents such as hydrogen atom, monomer subunit etc. For example, low density polythene.

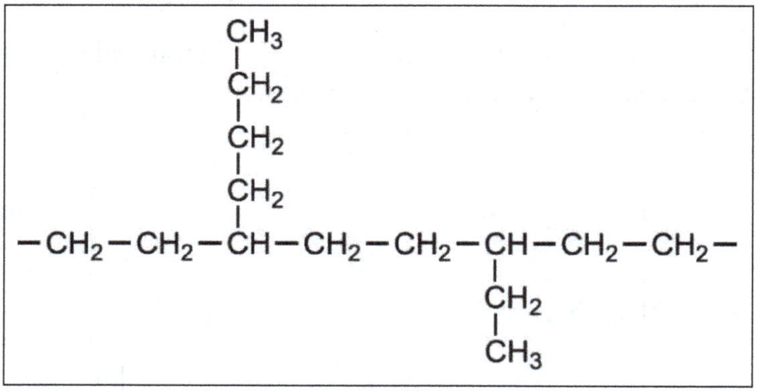

Figure: Low density polythene.

- Network Polymers are formed by combination of bi-functional and tri-functional monomers or cross-linking between different polymer chains. They are formed by the formation of covalent bonds between various linear polymers. For example, bakelite, melamine etc.

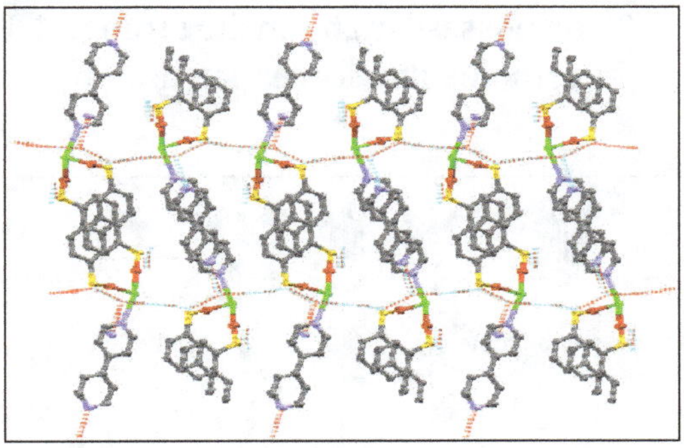

Figure: 3D structure of network polymers.

Classification based on the Mode of Polymerization

- Addition Polymers are formed by the repeated addition of monomers containing double or triple bonds. When the addition occurs via a same monomeric unit, the monomers formed are known as homo-polymers. When different units of monomers are used, the polymer formed is known as copolymer. For example, nitrile rubber.

```
(---(CH2-CH=CH-CH2) ---(CH2--CH) ---)n
                             1
                             CN
```

Figure: Structure of nitrile rubber.

- Condensation Polymers are formed by the condensation of different monomeric units. It involves the elimination of water, alcohol, etc. For example, formation of nylon 6,6.

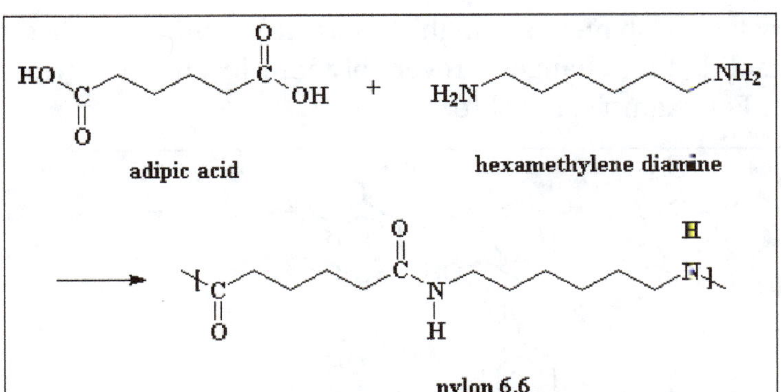

Figure: Structure of Nylon-6,6.

Classification based on Molecular Forces

- Fibers are synthetic as well as natural polymers. Synthetic fibers are cheap and easy to made whereas natural fibers are more comfortable for use as compared to synthetic fibers. They possess high tensile strength. They are thread forming solids with strong intermolecular forces such as hydrogen bonds. For example, Terylene.

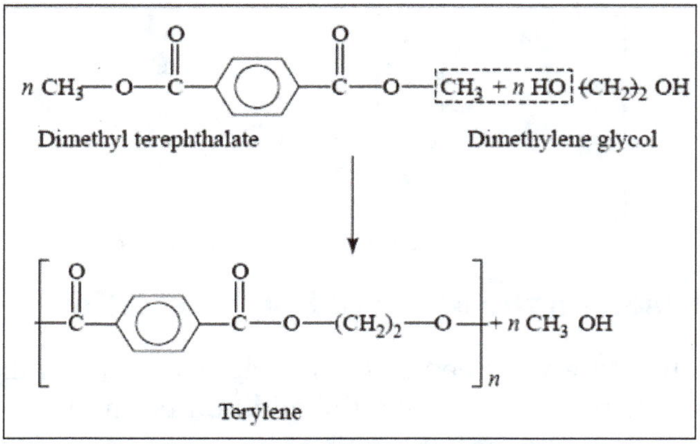

Figure: Structure of Terylene.

- Elastomers are polymers that possess both viscosity and elasticity. For example, rubber, are elastic solids. The polymer is held by weak intermolecular forces. They have the property to retract to its original position after being stretched.

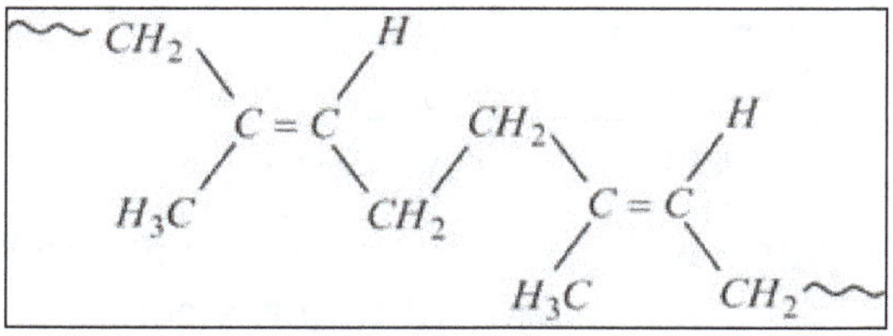

Figure: Structure of rubber.

- Thermosetting Polymers are highly cross-linked and branched polymers. They are soft solids that changes irreversibly into insoluble polymer. They are not reusable. For example, bakelite.

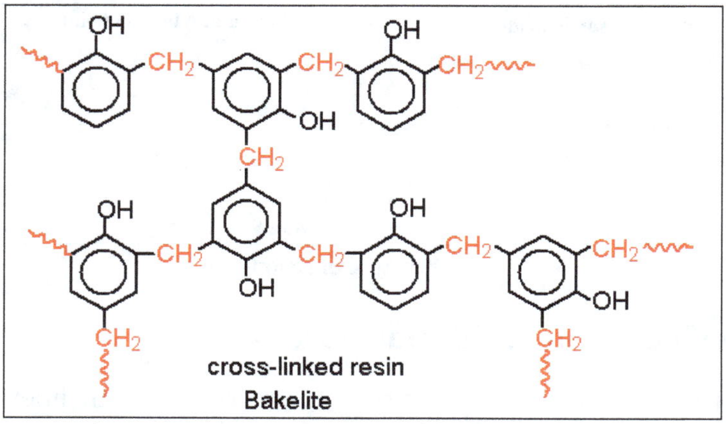

Figure: Structure of bakelite.

- Thermoplastic Polymers are linear or slightly branched polymers which becomes soft on heating while becomes hard on cooling. They are moldable above a specific temperature. For example, polythene, polystyrene etc.

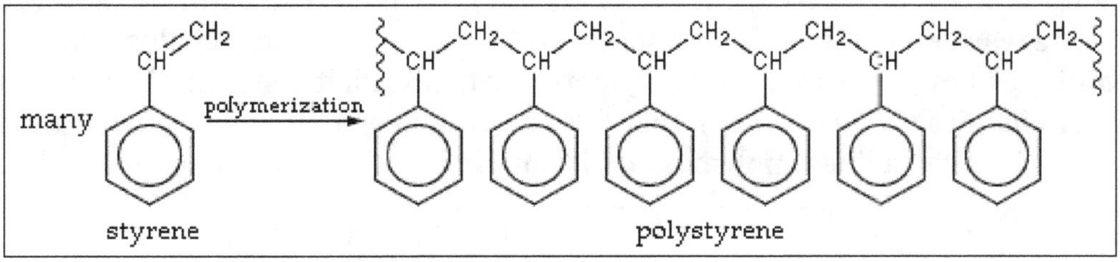

Figure: Structure of polystyrene.

Structure of Polymers

Polymers are very useful materials because their structures can be altered and tailored to produce materials:

- With a range of mechanical properties.
- In a wide spectrum of colors and 3) with different transparent properties.

Mers

A polymer is composed of many simple molecules that are repeating structural units called monomers. A single polymer molecule may consist of hundreds to a million monomers and may have a linear, branched, or network structure. Covalent bonds hold the atoms in the polymer molecules together and secondary bonds then hold groups of polymer chains together to form the polymeric material. Copolymers are polymers composed of two or more different types of monomers.

Polymer Chains (Thermoplastics and Thermosets)

A polymer is an organic material and the backbone of every organic material is a chain of carbon atoms. The carbon atom has four electrons in the outer shell. Each of these valence electrons can form a covalent bond to another carbon atom or to a foreign atom. The key to the polymer structure is that two carbon atoms can have up to three common bonds and still bond with other atoms. The elements found most frequently in polymers and their valence numbers are: H, F, Cl, Bf, and I with 1 valence electron; O and S with 2 valence electrons; n with 3 valence electrons and C and Si with 4 valence electrons.

The ability for molecules to form long chains is a vital to producing polymers. Consider the material polyethylene, which is made from ethane gas, C_2H_6. Ethane gas has a two carbon atoms in the chain and each of the two carbon atoms share two valence electrons with the other. If two molecules of ethane are brought together, one of the carbon bonds in each molecule can be broken and the two molecules can be joined with a carbon to carbon bond. After the two mers are joined, there are still two free valence electrons at each end of the chain for joining other mers or polymer chains. The process can continue liking more mers and polymers together until it is stopped by the addition of anther chemical (a terminator), that fills the available bond at each end of the molecule. This is called a linear polymer and is building block for thermoplastic polymers.

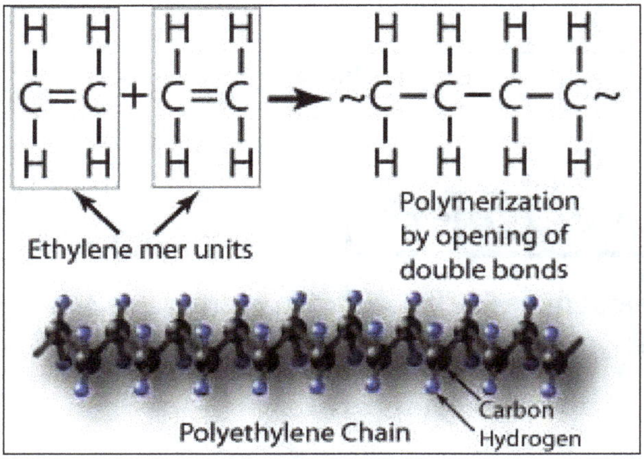

The polymer chain is often shown in two dimensions, but it should be noted that they have a three dimensional structure. Each bond is at 109° to the next and, therefore, the carbon backbone extends through space like a twisted chain of TinkerToys. When stress is applied, these chains stretch and the elongation of polymers can be thousands of times greater than it is in crystalline structures.

The length of the polymer chain is very important. As the number of carbon atoms in the chain is increased to beyond several hundred, the material will pass through the liquid state and become a waxy solid. When the number of carbon atoms in the chain is over 1,000, the solid material polyethylene, with its characteristics of strength, flexibility and toughness, is obtained. The change in state occurs because as the length of the molecules increases, the total binding forces between molecules also increases.

It should also be noted that the molecules are not generally straight but are a tangled mass. Thermoplastic materials, such as polyethylene, can be pictured as a mass of intertwined worms randomly thrown into a pail. The binding forces are the result of van der Waals forces between molecules and mechanical entanglement between the chains. When thermoplastics are heated, there is more molecular movement and the bonds between molecules can be easily broken. This is why thermoplastic materials can be re-melted.

There is another group of polymers in which a single large network, instead of many molecules is formed during polymerization. Since polymerization is initially accomplished by heating the raw materials and bringing them together, this group is called thermosetting polymers or plastics. For this type of network structure to form, the mers must have more than two places for boning to occur; otherwise, only a linear structure is possible. These chains form jointed structures and rings, and may fold back and forth to take on a partially crystalline structure.

Figure: Different Types of Carbon Backbone Structures.

Since these materials are essentially comprised of one giant molecule, there is no movement between molecules once the mass has set. Thermosetting polymers are more rigid and generally have higher strength than thermoplastic polymers. Also, since there is no opportunity for motion between molecules in a thermosetting polymer, they will not become plastic when heated.

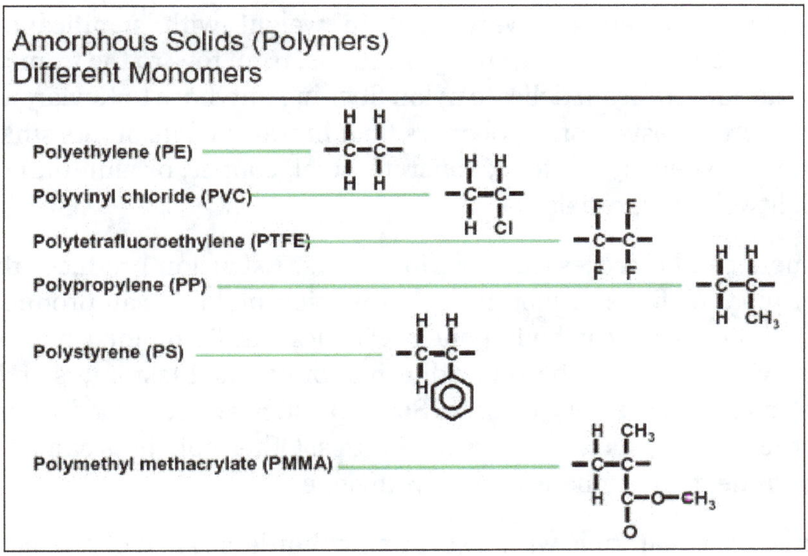

Characteristics of Polymers

The majority of manufactured polymers are thermoplastic, meaning that once the polymer is formed it can be heated and reformed over and over again. This property allows for easy processing and facilitates recycling. The other group, the thermosets, cannot be re-melted. Once these polymers are formed, reheating will cause the material to ultimately degrade, but not melt.

Every polymer has very distinct characteristics, but most polymers have the following general attributes:

- Polymers can be very resistant to chemicals. Consider all the cleaning fluids in your home that are packaged in plastic. Reading the warning labels that describe what happens when the chemical comes in contact with skin or eyes or is ingested will emphasize the need for chemical resistance in the plastic packaging. While solvents easily dissolve some plastics, other plastics provide safe, non-breakable packages for aggressive solvents.

- Polymers can be both thermal and electrical insulators. A walk through your house will reinforce this concept, as you consider all the appliances, cords, electrical outlets and wiring that are made or covered with polymeric materials. Thermal resistance is evident in the kitchen with pot and pan handles made of polymers, the coffee pot handles, the foam core of refrigerators and freezers, insulated cups, coolers, and microwave cookware. The thermal underwear that many skiers wear is made of polypropylene and the fiberfill in winter jackets is acrylic and polyester.

- Generally, polymers are very light in weight with significant degrees of strength. Consider the range of applications, from toys to the frame structure of space stations, or from delicate nylon fiber in pantyhose to Kevlar, which is used in bulletproof vests. Some polymers float in water while others sink. But, compared to the density of stone, concrete, steel, copper, or aluminum, all plastics are lightweight materials.

- Polymers can be processed in various ways. Extrusion produces thin fibers or heavy pipes or films or food bottles. Injection molding can produce very intricate parts or large car body panels. Plastics can be molded into drums or be mixed with solvents to become adhesives or paints. Elastomers and some plastics stretch and are very flexible. Some plastics are stretched in processing to hold their shape, such as soft drink bottles. Other polymers can be foamed like polystyrene, polyurethane and polyethylene.

- Polymers are materials with a seemingly limitless range of characteristics and colors. Polymers have many inherent properties that can be further enhanced by a wide range of additives to broaden their uses and applications. Polymers can be made to mimic cotton, silk, and wool fibers; porcelain and marble; and aluminum and zinc. Polymers can also make possible products that do not readily come from the natural world, such as clear sheets and flexible films.

- Polymers are usually made of petroleum, but not always. Many polymers are made of repeat units derived from natural gas or coal or crude oil. But building block repeat units can sometimes be made from renewable materials such as polylactic acid from corn or cellulosic from cotton linters. Some plastics have always been made from renewable materials such as cellulose acetate used for screwdriver handles and gift ribbon. When the building blocks can be made more economically from renewable materials than from fossil fuels, either old plastics find new raw materials or new plastics are introduced.

- Polymers can be used to make items that have no alternatives from other materials. Polymers can be made into clear, waterproof films. PVC is used to make medical tubing and blood bags that extend the shelf life of blood and blood products. PVC safely delivers flammable oxygen in non-burning flexible tubing. And anti-thrombogenic material, such as heparin, can be incorporated into flexible PVC catheters for open heart surgery, dialysis, and blood collection. Many medical devices rely on polymers to permit effective functioning.

References

- Definition-of-polymer-605912: thoughtco.com, Retrieved 21, July 2020
- Polymers-60682: livescience.com, Retrieved 10, March 2020
- Biopolymers-application-in-nanoscience-and-nanotechnology, recent-advances-in-biopolymers: intechopen.com, Retrieved 29, August 2020
- Classification-of-polymers, polymers, inorganic-chemistry: askiitians.com, Retrieved 05, February 2020
- Polymer, structure, materials, community college: nde-ed.org, Retrieved 17, July 2020

2

Properties of Polymers

The properties of polymers depend on their structure. Mechanical properties of polymers dictate how polymers behave on a macroscopic scale. Transport properties describe how rapidly molecules move through the polymer matrix. Various properties of polymers like molecular properties, transport properties, mechanical properties, optical properties, electrical properties and phase behavior etc. are discussed in detail in this chapter.

Molecular Properties of Polymers

Almost all properties of a polymer depend on the molecular weight and its distribution, that is, low and high molecular weight polymers will have vastly different mechanical and thermo-physical properties. For example, polymeric materials consisting of a rather small number of repeat units (often called *oligomers*) will be soft or rubbery solids or low viscous liquids and will possess little or no strength whereas polymeric materials of high molecular weight will be solid, and have much improved mechanical properties. One usually observes a steep rise in the mechanical and visco-elastic properties with increasing molecular weight until a certain molecular weight is reached, beyond which the properties are nearly independent of the chain length, that is, all properties will eventually reach an asymptotic value. It was found, that the molecular weight dependence of several thermodynamical properties can be described by an expression of the form.

$$X(M_n) = X_\infty - \frac{B}{M_n}$$

where X is a certain thermodynamical property, X_∞ is its asymptotic value at infinitely high molecular weight, and B is a constant.

Some polymer properties that are very sensitive to the molecular weight include softening and melting point, glass transition temperature, solution and melt viscosity, solubility, mechanical strength and moduli.

As an example, in the figure the glass transition temperature (T_g) as a function of molecular weight is plotted for polystyrene. The T_g values have been predicted with the Flory-Fox equation.

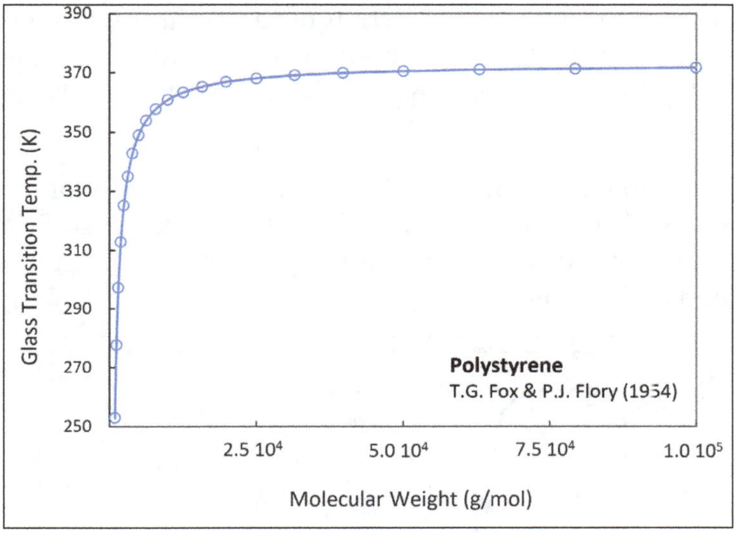

$$Tg(k) = T_{g,\infty} - \frac{K}{M_n} = 373K - \frac{12 \cdot 10^4}{M_n}$$

Where $T_{g,\infty}$ is the asymptotic glass transition temperature at very high molecular weight.

Not only the average molecular weight but also its distribution has a noticeable effect on the physical, mechanical, and processing properties. In most cases, a narrow molecular weight distribution will yield better mechanical properties than a broad distribution because the low molecular weight portion will act as a plasticizer and soften the polymeric material, whereas an exceedingly high molecular weight portion will make processing of the polymer resin very difficult due to its disproportionately high contribution to the melt viscosity.

Mechanical Properties of Polymers

The term mechanical properties are commonly used to denote stress-strain relationships for polymer systems. Unlike many more familiar materials where these relationships depend essentially only on temperature, in polymeric systems time dependence is also of importance. This time dependence necessitates very careful definitions of parameters such as moduli and compliances which result from experiments involving discontinuous stress or strain levels. In addition, the time dependence may be explored using oscillatory perturbations as is done when investigating dynamic mechanical properties or dielectric relaxation in polymers.

An experimental “master curve” depicts the behavior of a polymeric system at any temperature over a broad spectrum of time. On the other hand, mechanical behavior may be presented isochronally as a function of temperature; one usually observes four distinct regions of viscoelastic behavior in such a curve for a linear amorphous polymer and each of these regions may be related to specific types of molecular motion.

Voigt and Maxwell elements are frequently used in the treatment of mechanical properties. It is well known that elements such as these do not possess sufficient flexibility to mimic experimentally observed behavior with high fidelity, but combinations of these elements are found to do a much better job.

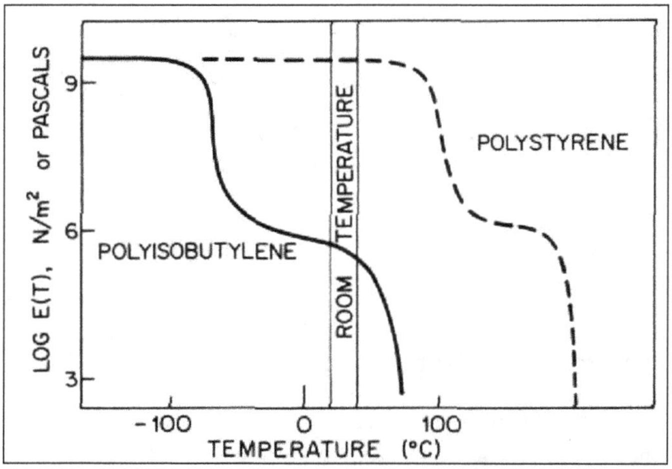

Figure: Schematic Modulus-temperature curves for a rubber (polyisobutylene) and for a plastic (polystyrene).

Modulus-temperature Behavior of Polymers

Although we recognize that time plays a very important role in the determination of mechanical properties of polymers, for the moment, let us ignore this fact. Here, we will concentrate on the influence of temperature in determining mechanical properties of polymers.

Let us start by considering a simple tensile experiment carried out on a strip of polyisobutylene (PIB). For the sake of concreteness, assume that the strip is 2 cm long and has a cross sectional area of 0.3 cm^2. At this level of approximation, a tensile experiment simply consists of hanging a weight on this strip of PIB and observing the change in sample length {Al). Under these conditions, the tensile modulus is defined as,

$$E = \sigma / \varepsilon = (f / A) / \Delta l / l$$

Here, E is the tensile modulus; a, the tensile stress defined as the force applied to the system divided by the cross sectional area, A; and e, the tensile strain which is equal to the change in sample length divided by the original sample length l. alternatively, one might define the tensile compliance.

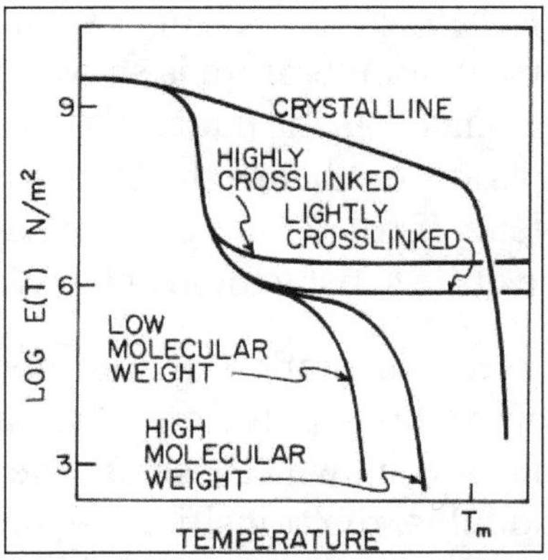

Figure: Changes in the schematic representations of modulus-temperature behavior of a polymer caused by molecular weight variation, cross-linking and crystallization.

$$D = \frac{\varepsilon}{\sigma} = \frac{1}{E}$$

Let us imagine an experiment in which the strip of PIB is cooled, perhaps in liquid nitrogen, to a rather low temperature; the strip is subjected to a force and the resultant sample extension is observed. At this low temperature, the resistance to elongation of a material like PIB is substantial and the calculated modulus, shown in figure, is correspondingly high with a value in the range of 3 X 10^9 Newtons/meter^2 (3 X 10^{10}Dyne/cm^2). Such high modulus values are characteristic of glassy or plastic materials.

The sample is next warmed several degrees and an identical experiment carried out; similar results are observed until, in the neighborhood of —70°C for PIB, a dramatic decrease in modulus by approximately a factor of one thousand is observed. Now the modulus has a value in the range of 10^6 N/m^2 which is characteristic of rubbery materials. Thus, near —70°C, the properties of polyisobutylene change from those of a glass to those of a rubber and, it is said that polyisobutylene has undergone its "glass transition" near this temperature.

Upon further heating, the modulus remains reasonably constant until considerably higher temperatures are reached. At these higher temperatures, another drastic decrease in modulus is observed for linear amorphous materials. In this range, the sample actually becomes liquid-like and does not support any stress, not even the stress generated by its own weight. The overall modulus versus temperature behavior for PIB is schematically depicted in Figure above.

Rather than polyisobutylene, one might have chosen another common polymer such as polystyrene (PS) for this series of experiments. With PS, the results would have been somewhat different. Here, at — 70°C no drastic decrease in modulus is observed. Rather, glassy behavior persists to temperatures near 100°C. At that point, the change from glassy to rubbery-like behavior is observed; this is consistent with the observation that the glass transition temperature for polystyrene is very close to 100°C. At higher temperatures still, the modulus behavior of PS mimics that of PIB except for a temperature shift. A band indicating room temperature is shown. It is clear in this representation why polystyrene is used to make plastic materials, while polyisobutylene is used to fabricate rubbers. Room temperature is below the glass transition range of polystyrene yet above that of polyisobutylene. If room temperature happened to be at 140°C, polystyrene could be used to make quite satisfactory rubber bands.

In Figure below, a schematic representation of the modulus temperature behavior which is common to linear amorphous polymers is presented. Here, the four major regions of viscoelastic behavior are shown. These are the glassy region, characterized by a very high value of the modulus, the transition region from glassy to rubbery-like consistency, the rubbery plateau at much lower modulus values than those characteristic of glasses and finally, at high temperatures, a flow region.

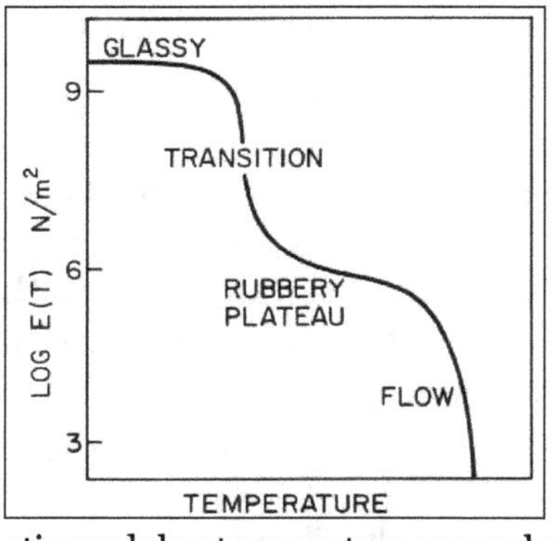

Figure: Schematic modulus-temperature curve showing the four regions of viscoelastic behavior.

The behavior of polymer molecules in each of these regions is very different. In the glassy domain the relatively low values of temperature result in limited extents of molecular motion. Here, short range motions of side chains on the polymer backbone or perhaps limited backbone motions involving short segments of chains take place. As the temperature increases, a situation is reached where substantial segments of polymer chains have enough energy to surmount local barriers which hinder molecular motion and these segments begin to move. In this range, motions on a scale of perhaps ten monomer units might occur. This is the transition zone. At higher temperatures still, even longer range motion involving larger segments of chains become possible? However, due to the fact that long polymer molecules are entangled, whole scale translational motions of polymer chains are forbidden. This is the molecular interpretation of behavior in the rubbery plateau region. At higher temperatures still, molecular motion is so extreme that even chain entanglements are no longer effective in restricting flow of one molecule past another and a liquid-like regime is realized. Thus, with increasing temperature, the extent of molecular motion in polymeric systems increases giving rise to the four domains of viscoelastic behavior discussed.

With this simple interpretation of molecular behavior, it is possible to rationalize and often predict modulus-temperature curves for a variety of circumstances, a very valuable asset. This is done with reference to a linear amorphous polymer of moderate molecular weight. First, let us consider the influence on the modulus temperature behavior associated with increasing molecular weight. Suppose, for example, that the molecular weight of our sample changed from M_n = 200,000 g/mole to 500,000 g/mole. For such high molecular weight materials, the fraction of chain ends is so small as to be negligible. Thus, for the short range motions involved in the glassy and transition regions, chain length is rather unimportant and such a change in molecular weight has little effect. However, since longer chains have more entanglements on average than shorter ones, one might anticipate of the flow regime in the case of the higher molecular weight polymer.

Cross-linking, that is covalent bonding of one chain to another, results in permanent restrictions to flow which do not slacken with increasing temperature. Thus, a cross-linked polymer does not exhibit a flow regime. In samples which are cross-linked, the modulus depends on the number of chains which must be deformed to elongate a sample. In more highly cross-linked systems the density of network chains is higher and thus the modulus increases with increasing cross-link density. This, too, is shown in the figure.

Often, polymer chains have such regularity that crystallization is possible. Usually, in bulk samples, domains of crystalline material are scattered randomly throughout the sample. As can be imagined, these crystals restrict flow of one chain by another in much the same way as does a cross-link. Thus, at temperatures below their crystal melting points, semi-crystalline polymers behave like cross-linked materials (the crystal melting temperature is usually considerably above the glass transition temperature). The behavior shown for the crystalline material is that of a highly crystalline polymer with a melting temperature (T_m) indicated. Below Tm, the numerous crystals restrict motion and result in a high modulus material. Above Tm, the crystals melt leaving the chains free to flow at temperatures substantially above T_g for the particular polymer.

Many often more complicated types of polymeric behavior can be understood in terms of these simple molecular ideas.

Time Dependence

In addition to the temperature sensitivity of the mechanical properties of polymers, there is a marked dependence on time. For example, in the modulus determinations, the results would have varied substantially if strain measurements had been made 10 sec after the imposition of the stress or 10 min after the stress application. Furthermore, the particular method by which the measurement is carried out is also important. The experiments discussed above are called creep experiments. The sample is subjected to a constant stress and the strain increases with time. Suppose that a creep experiment had been carried out on a particular sample and, after a particular amount of time, say 1 min, a certain strain was observed. Given the constant stress value and the strain observed after 1 min, a 1 min modulus value could be calculated. On the other hand, suppose that the modulus measurement was carried out in an alternate way. Let us imagine that the sample is rapidly elongated to the strain observed after 1 min in the previous experiment (theoretically the deformation is instantaneous) and that this strain is then held constant. Initially, the stress necessary to produce the rapid strain is high but this stress decays with time as relaxation occurs. Such a constant strain experiment is called a stress relaxation experiment and, here to a 1-min modulus can be calculated from the stress and strain values after 1 min. The modulus after 1 min measured in the creep experiment can be substantially different (an order of magnitude) from that measured in the stress relaxation experiment. Such differences in values result from the time dependence of the mechanical properties of polymers. Therefore, it is very important to define parameters which characterize mechanical properties of polymers in a precise fashion.

The tensile creep compliance, D(t), is defined as,

$$D(t) = \varepsilon(t)/\sigma_o$$

Where $\varepsilon(t)$ is the tensile strain at time t after the imposition of an instantaneous stress σ_0 at t = 0. Similarly, the tensile stress relaxation modulus, E(t) may be defined as:

$$E(t) = \sigma(t)/\varepsilon_o$$

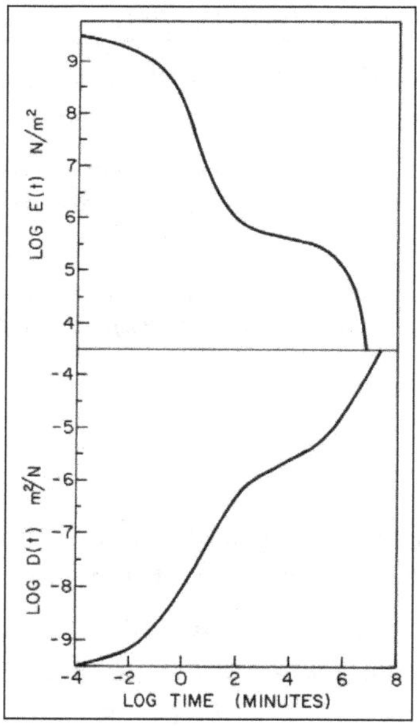

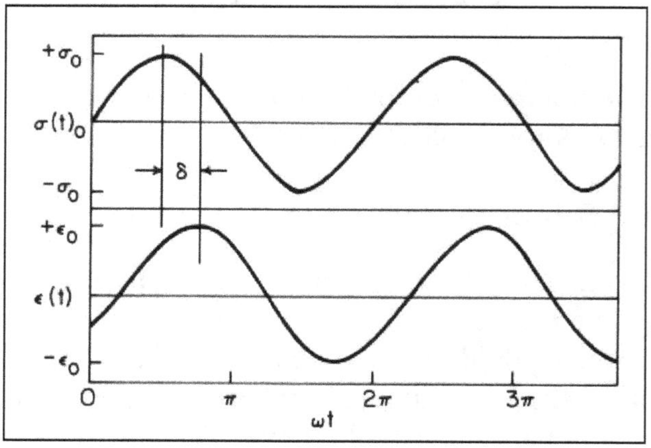

where, $\sigma(t)$ is the stress at time t necessary to maintain a constant strain ε_0 which was applied to the sample at t= 0. These are but two of the many possible definitions of moduli and compliances for materials with time-dependent mechanical properties. Here it is clear that the time-dependent tensile compliance and the time-dependent stress relaxation modulus are not simply reciprocally related as is the case in the absence of time dependence. Rather, it can be shown that their relationship is given by the convolution integral:

$$t = \int_0^t E(a)D(t-a)da$$

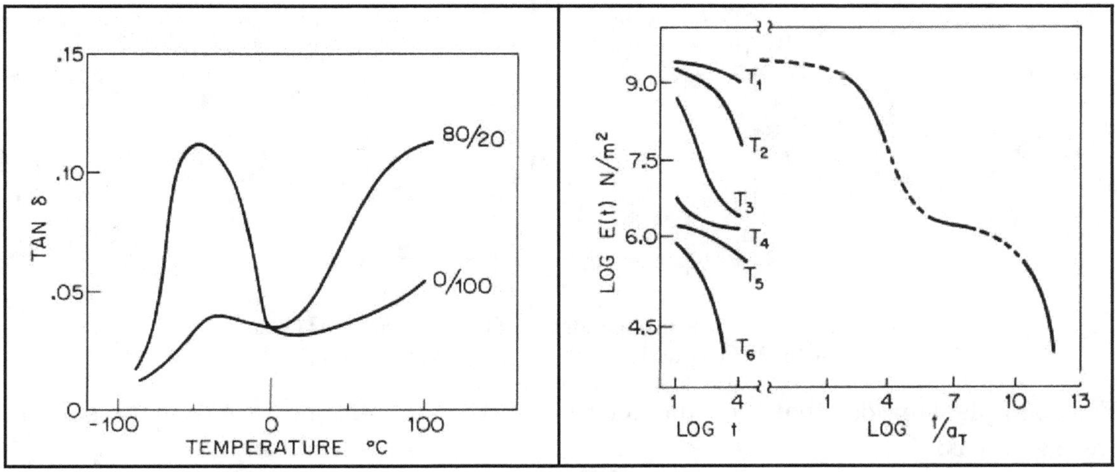

In figure a schematic representation of E(t) and D(t) for the same material at the same temperature is presented. Such curves, often referred to as "master curves," represent the time dependence of a polymeric system at fixed temperature over a very broad time scale.

An exact reciprocal relationship would result in the stress relaxation and creep master curves being mirror images of each other. Although this is approximately true, the exact relationship between the curves is more complex, as given in equation $t = \int_0^t E(a)D(t-a)da$, causing the creep master curve to be somewhat broader than that observed in stress relaxation.

The general shape of the stress relaxation master curve is reasonably similar to that of a modulus-temperature curve. (From this point forward, we will consider that modulus temperature curves are prepared with modulus values measured at a constant time period.) Not surprisingly, the molecular interpretations of these various regions are also similar to those given above. For example, in the glassy region time is too short to allow substantial rearrangements of large molecular entities and the resulting modulus is, therefore, high. On the other hand, over a very protracted time scale even at the same low temperature, longer range molecular motions become possible with the corresponding decrease in modulus observed.

The measurement of dynamic mechanical properties offers a convenient way to assess time dependence and is commonly done in many laboratories. Here, rather than subjecting the material to a step change in stress or strain, an oscillatory perturbation is applied to the sample. The corresponding oscillatory response is then monitored and analyzed.

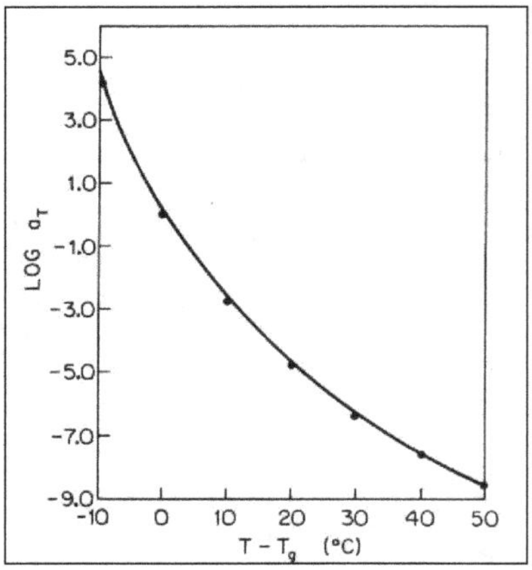

Figure: Temperature dependence of the shift factor as given by the WLF equation with C_i = 17.4 and C_s = 51,6°K.

For example, suppose that the time dependent stress imposed on a polymeric sample were given as:

$$\sigma(t) = \sigma_o \sin(\omega t)$$

This function depicts a stress which oscillates from $+\sigma_o$ to $-\sigma_o$ with frequency ω. (Here it is assumed that the oscillation has been imposed for a long time such that initial transient effects are no longer important.) Since the material subjected to this stress pattern has time dependent mechanical properties, it is found that the resulting strain, although sinusoidal, will lag the stress by some amount of time. Thus, the strain maximum can come at a later time than the stress maximum. For a material which is highly elastic, the stress and strain maxima will occur at essentially the same point. For a highly viscous material, the strain maximum will lag the stress maximum by approximately one quarter cycle. A very convenient method to represent dynamic mechanical behavior of polymers involves rotating vectors. One vector represents the stress and the other the strain. These vectors rotate with frequency cu and their projections onto any line represent the data. The phase angle δ, between these two vectors will vary with frequency and temperature and reflects the time and temperature dependence of the mechanical properties of the polymer under study. Figure above is a plot of tangent δ as a function of temperature for two polymers under conditions of constant frequency (200 Hz).

Time-Temperature Correspondence

The time scale over which creep and stress relaxation master curves are presented is exceedingly broad extending over some twelve to fifteen decades of time. Since it is impossible to observe the behavior directly over such broad time scales, a technique has been developed which allows the preparation of master curves from experimental data. In Figure below on the left, data is represented over the reasonably accessible time scale range of two to three decades. Here, the tensile stress relaxation modulus is depicted at several temperatures. With the help of the time-temperature correspondence principle, such experimental data measured on accessible time scales at various temperatures can be used to prepare a master curve at a fixed temperature over a very broad time scale.

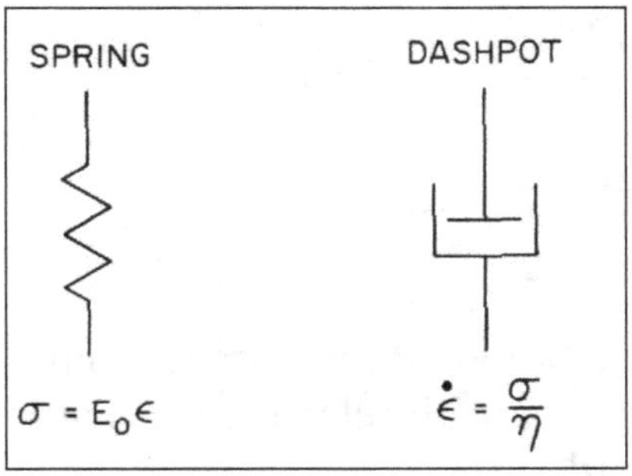

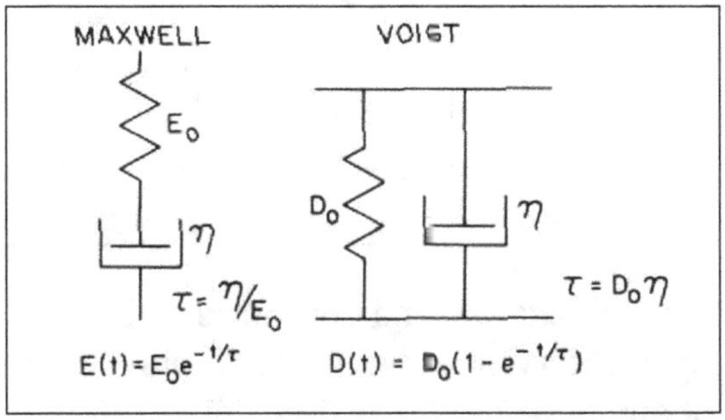

The time-temperature correspondence principle asserts that if one were interested in knowing the stress relaxation modulus behavior of the material represented in figure above at the temperature T_1 at times longer than 104 sec, either of two procedures could be used. Firstly, the direct evaluation of the modulus at temperature T_t for periods longer than 104 sec could be measured. Alternatively, the experimental temperature might be raised to a somewhat higher temperature, 7V If the data measured at T% is then simply shifted horizontally, a portion of the To curve will superimpose itself upon the previously measured T_i curve. The time-temperature correspondence principle asserts that the extension of the T_2 curve beyond the T_1 curve is exactly the behavior which would be measured at the lower temperature T in a stress relaxation experiment of longer duration. The time-temperature correspondence principle may be applied to a series of overlapping log modulus versus log time curves at various temperatures to yield a stress relaxation master curve. (Small vertical adjustments are also sometimes necessary.) In this figure, the temperature T_2 has been chosen as the reference temperature for the preparation of the master curve. Thus, data measured at temperatures below the reference temperature are shifted to shorter times while data measured at temperatures above the reference temperature are shifted to longer times. This shifting may be represented as a division of the time scale by a factor α_T as is shown in the figure. The temperature dependence of the shift factors is shown in Figure above. Here, log α_T is plotted versus temperature. For this figure, it is assumed that the glass transition temperature, T_g, was chosen as the reference temperature. The temperature dependence of the shift factor can be expressed analytically in terms of the WLF equation:

$$\log a_T = \frac{-C_1\left(T - T_g\right)}{C_2 + T - T_2}$$

where, C_1 and C_2 are constants. Although it was originally thought that C_i and C_2 were universal constants, it is agreed that they vary from polymer to polymer. The "universal" values for C_i at C_2 were 17.4 and 51.6°, respectively.

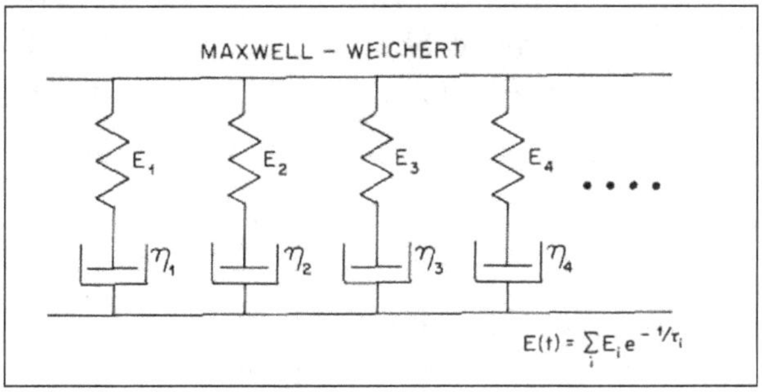

The stress relaxation master curve, the stress relaxation modulus versus temperature curve and the WLF equation for a particular polymer are all related such that knowledge of any two allows the determination of the third.

Thermal Properties of Polymers

Polymer Glass Transition

It is important to understand the concept of the glass transition temperature, T_g. As the temperature of a polymer drops below T_g, it behaves in an increasingly brittle manner. As the temperature rises above the T_g, the polymer becomes more rubber-like. Thus, knowledge of T_g is essential in the selection of materials for various applications. In general, values of T_g well below room temperature define the domain of elastomers and values above room temperature define rigid, structural polymers.

This behavior can be understood in terms of the structure of glassy materials which are formed typically by substances containing long chains, networks of linked atoms or those that possess a complex molecular structure. Normally such materials have a high viscosity in the liquid state. When rapid cooling occurs to a temperature at which the crystalline state is expected to be the more stable, molecular movement is too sluggish or the geometry too awkward to take up a crystalline conformation. Therefore the random arrangement characteristic of the liquid persists down to temperatures at which the viscosity is so high that the material is considered to be solid. The term glassy has come to be synonymous with a persistent non-equilibrium state. In fact, a path to the state of lowest energy might not be available.

To become more quantitative about the characterization of the liquid-glass transition phenomenon and T_g, we note that in cooling an amorphous material from the liquid state, there is no abrupt change in volume such as occurs in the case of cooling of a crystalline material through its freezing point, T_f. Instead, at the glass transition temperature, T_g, there is a change in slope of the curve of specific volume vs. temperature, moving from a low value in the glassy state to a higher value in the rubbery state over a range of temperatures. This comparison between a crystalline material (1) and an amorphous material (2) is illustrated in the figure below. Note that the intersections of the two straight line segments of curve (2) define the quantity T_g.

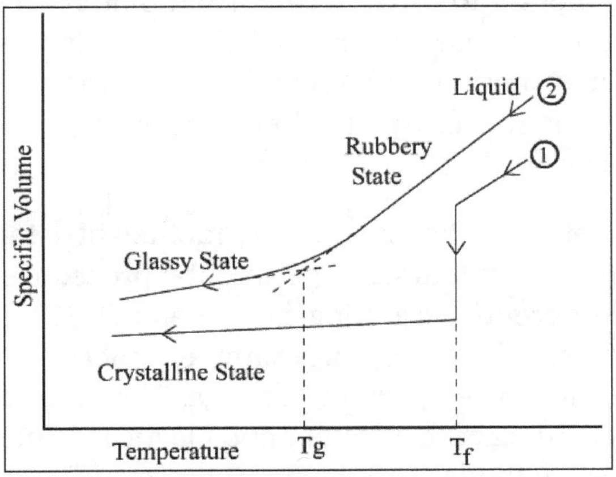

The specific volume measurements shown here, made on an amorphous polymer (2), are carried out in a dilatometer at a slow heating rate. In this apparatus, a sample is placed in a glass bulb and a confining liquid, usually mercury, is introduced into the bulb so that the liquid surrounds the sample and extends partway up a narrow bore glass capillary tube. A capillary tube is used so that relatively small changes in polymer volume caused by changing the temperature produce easily measured changes in the height of the mercury in the capillary.

The determination of T_g for amorphous materials, including polymers as mentioned above, by dilatometric methods (as well as by other methods) are found to be rate dependent. This is schematically illustrated in the figure below, again representing an amorphous polymer, where the higher value, T_{g2}, is obtained with a substantially higher cooling rate than for T_{g1}.

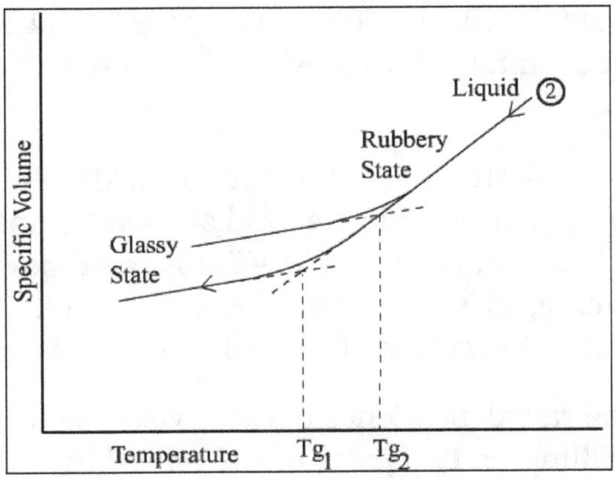

We can understand this rate dependence in terms of intermolecular relaxation processes. Since a glass is not an equilibrium phase, its properties will exhibit a time dependence, or physical aging. The primary portion of the relaxation behavior governing the glass transition in polymers can be related to their tangled chain structure where cooperative molecular motion is required for internal readjustments. At temperatures well above T_g, 10 to 50 repeat units of the polymer backbone are relatively free to move in cooperative thermal motion to provide conformational rearrangement of the backbone. Below T_g, the motion of these individual chains segments becomes frozen with only small scale molecular motion remaining, involving individual or small groups of atoms. Thus a rapid cooling rate or "quench" takes rubbery material into glassy behavior at higher temperatures (higher T_g).

While the dilatometer method is the more precise method of determining the glass transition temperature, it is a rather tedious experimental procedure and measurements of T_g are often made in a differential scanning calorimeter (DSC). In this instrument, the heat flow into or out of a small (10 – 20 mg) sample is measured as the sample is subjected to a programmed linear temperature change. There are other methods of measurement such as density, dielectric constant and elastic modulus which are treated in texts on polymers. These methods are, of course, also rate dependent.

T_g and Mechanical Properties

Another important property of polymers, also strongly dependent on their temperatures, is their response to the application of a force, as indicated by two main types of behavior: elastic and plastic. Elastic materials will return to their original shape once the force is removed. Plastic materials will not regain their shape. In plastic materials, flow is occurring, much like a highly viscous liquid. Most materials demonstrate a combination of elastic and plastic behavior, showing plastic behavior after the elastic limit has been exceeded.

Glass is one of the few completely elastic materials while it is below its T_g. It will remain elastic until it reaches its breaking point. The T_g of glass occurs between 510 and 560 degrees C, meaning that it will always be a brittle solid at room temperature. In comparison, polyvinyl chloride (PVC) has a T_g of 83 degrees C, making it good, for example, for cold water pipes, but unsuitable for hot water. PVC also will always be a brittle solid at room temperature.

Adding a small amount of plasticizer to PVC can lower the T_g to – 40 degrees C. This addition renders the PVC a soft, flexible material at room temperature, ideal for applications such as garden hoses. A plasticized PVC hose can, however, become stiff and brittle in winter. In this case, as in any other, the relation of the T_g to the ambient temperature is what determines the choice of a given material in a particular application.

A striking example of the rate dependence of these viscoelastic properties is furnished by Silly Putty. Slowly pulling on two parts of the Silly Putty stretches it apart until it very slowly separates. Placing the Silly Putty on a table and hitting it with a hammer will shatter it.

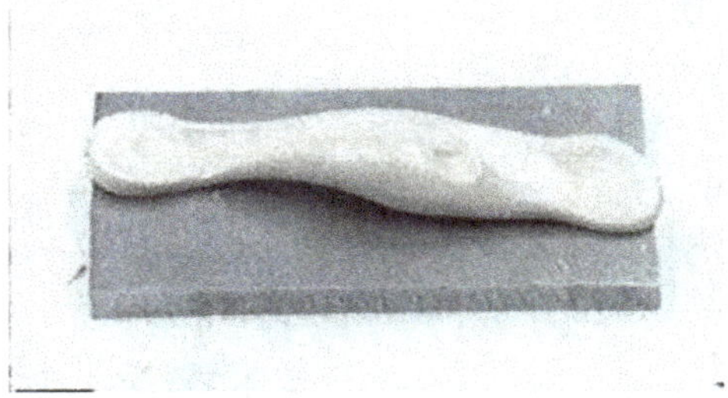

Slowly Deformed

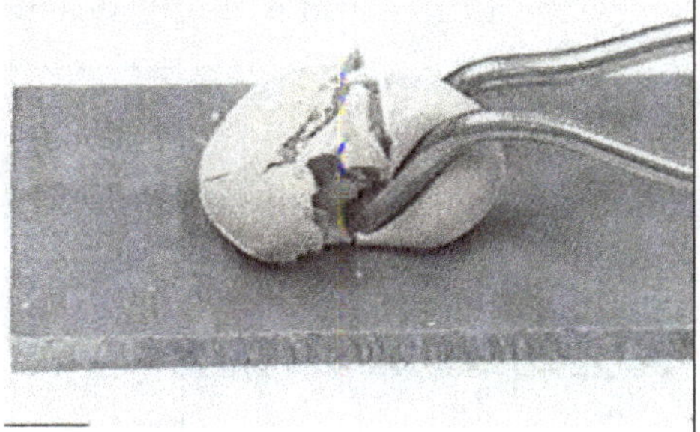

Rapidly Deformed

The images are representative of the behavior of a material above and below its glass transition temperature. The image on the (left) is Silly Putty that has been slowly stretched. The image on the (right) is Silly Putty which has been hit with a hammer. The speed of the hammer raised the rate of the application of the force and in turn raised the T_g. This caused the Silly Putty to react as if it were below its T_g and to shatter. Even though both reactions took place at the same ambient temperature, one reaction appeared to be above the effective T_g and the other appeared to be below.

Differential Scanning Calorimetry

In differential scanning Calorimetry (DSC), the thermal properties of a sample are compared against a standard reference material which has no transition in the temperature range of interest, such as powdered alumina. Each is contained in a small holder within an adiabatic enclosure as illustrated below.

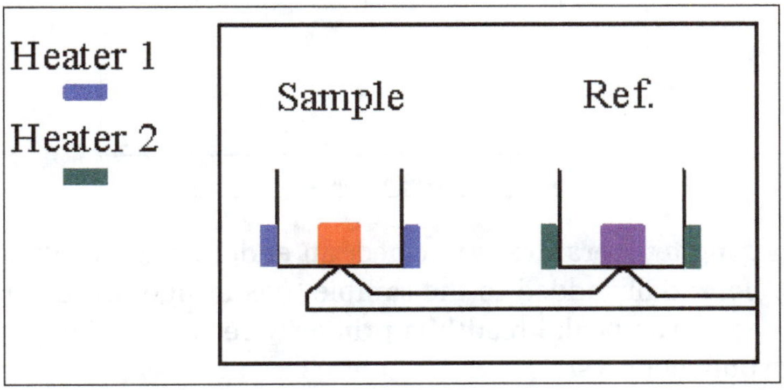

The temperature of each holder is monitored by a thermocouple and heat can be supplied electrically to each holder to keep the temperature of the two equal. A plot of the difference in energy supplied to the sample against the average temperature, as the latter is slowly increased through one or more thermal transitions of the sample yields important information about the transition, such as latent heat or a relatively abrupt change in heat capacity.

The glass transition process is illustrated in the figure below for a glassy polymer which does not crystallize and is being slowly heated from below T_g.

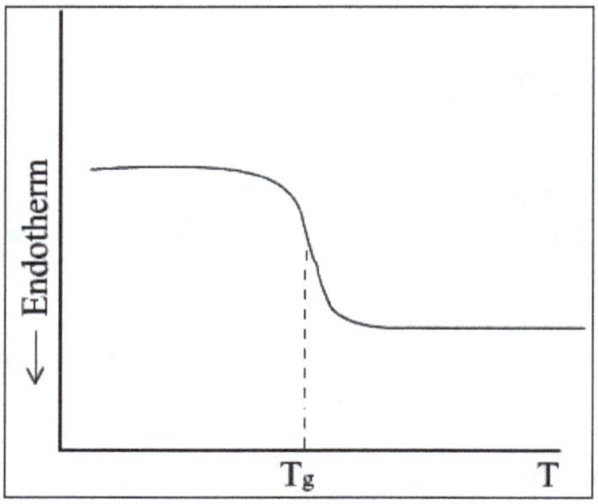

Here, the drop marked T_g at its midpoint represents the increase in energy supplied to the sample to maintain it at the same temperature as the reference material, due to the relatively rapid increase in the heat capacity of the sample as its temperature is raised through T_g. The addition of heat energy corresponds to this endothermal direction.

A melting process is also illustrated below for the case of a highly crystalline polymer which is slowly heated through its melting temperature.

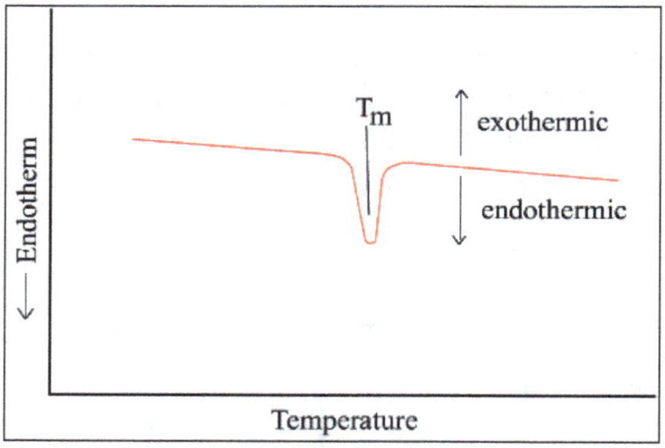

Again, as the melting temperature is reached, an endothermal peak appears because heat must be preferentially added to the sample to continue this essentially constant temperature process. The peak breadth is primarily related to the size and degree of perfection of the polymer crystals.

Note that if the process were reversed so that the sample was being cooled from the melt, the plot would be inverted. In that case, as both are being cooled by ambient conditions, even less heat would need to be supplied to the sample than to the reference material, in order that crystals can form. This corresponds to an exothermal process.

In the DSC plot, assume the temperature is slowly increased from this value below T_g to a value well above, thus allowing for significant increases in the chain mobility as temperatures above T_g are reached so that some crystallization can begin, well before the melting point is reached.

Optical Fiber

Plastic optical fibers (POFs) are a low-cost solution for low-speed, short distance applications in digital car networks, industrial networks, and home networks and appliances. The plastic optical fiber is made out of a plastic such as acrylic (PMMA) as the core material and fluorinated or perfluorinated polymers as the cladding materials.

Color Properties

Water Based Stains are similar to acid based stains in the sense that one can still achieve a translucent look like acid; some stains are able to achieve an opaque color and/or a translucent effect. The main difference is that acid stains react to the concrete and change the physical makeup of the concrete material, whereas water based stains are more of a "coating" that bonds with the concrete. There are many variations of water based stains that have come into the decorative concrete industry that perform in a number of different ways. Some are polymer based, acrylic and epoxy.

Concrete overlays date to the 1960s when chemical engineers from some of the larger, well known chemical companies began to experiment with acrylic paint resins as modifiers for cement and sand mixes. The result was a thin cementitious topping material that would adhere to concrete surfaces and provide a newly resurfaced coating to restore the worn surface. Concrete overlays lacked the long-term performance characteristics of acrylic resins. Acrylic resins provided good UV resistance, but lacked long term water resistance and adhesion characteristics needed to provide a long term and permanent solution.

Polymer cement overlays, consisting of a proprietary blend of Portland cements, various aggregates and polymer resins, were introduced over 20 years ago. The purpose of adding a polymer resin to the cement and aggregate is to greatly increase the performance characteristics and versatility of conventional cements, mortars and concrete materials. Unlike conventional cement and concrete mixes, polymer cement overlays can be applied thinly or thickly without fear of delamination or typical product failure. In addition, polymer cement overlays are much more resistant to damage from salt, petrochemicals, UV, harsh weather conditions and traffic wearing.

Originally intended for use as a thin surface restoration material for concrete substrates, polymer cement overlays were introduced into the architectural concrete and commercial flooring industries in the early 80s. Subsequently, its use in these industries has become standard. Polymer cement overlays are regarded as economical in providing long term, durable renovation without the need for costly and continuous repairs associated with deteriorating concrete surfaces.

Polymer cement overlays are used for interior and exterior applications ranging from:

- Skim coat/broom finish concrete resurfacing – Restore and protect damaged, pitted, flaking and stained concrete back to the look of a new concrete surface.
- Concrete regrading & leveling – Repair and level concrete surfaces that have settled.
- Existing substrate redecorating and renovating – Alter the appearance of existing concrete or wood substrates through applying "thin stamped" or "thin stained" overlays, creating new textures, colors and designs. For use on commercial or resident pool deck, this frequently takes the form of "splatter textures" or "knockdowns," in which polymer cement is applied to the existing concrete substrate in a moderately textured finish (average of 1/8" thickness) in various patterns. Often, the texture is knocked down with a trowel to slightly modify the appearance and feel of the finished application.

Physical Properties of Polymers

Physical properties of polymers include molecular weight, molar volume, density, degree of polymerization, crystallinity of material, and so on.

Degree of Polymerization and Molecular Weight

The degree of polymerization (DP)-n in a polymer molecule is defined as the number of repeating units in the polymer chain. For example,

$$-(-CH_2 - CH_2-)-n$$

The molecular weight of a polymer molecule is the product of the degree of polymerization and the molecular weight of the repeating unit. The polymer molecules are not identical but are a mixture of many species with different degrees of polymerization, that is, with different molecular weights. Therefore, in the case of polymers we talk about the average values of molecular weights.

Molecular Weight Averages

Suppose we have a set of values $\{x_1, x_2 ... x_n\}$ and the corresponding probability of occurrence is given by $\{P_1, P_2 ... P_n\}$, then the average value is defined as follows:

$$\sum_{i=0}^{\infty} P_i x_i$$

Number-Average Molecular Weight

If Ni is the number of polymer molecules having the molecular weight M_i, then the "number-average" probability of the given mass is given by:

$$P_i = \frac{N_i}{\sum_{j=o}^{\infty} N_j}$$

The number-average molecular weight is given by:

$$M_n = \sum_{i=o}^{\infty}\left[\frac{N_i}{\sum_{j=o}^{\infty} N_j}\right] Mi = \frac{\sum_{j=o}^{\infty} M_i N_i}{\sum_{j=o}^{\infty} N_j}$$

The physical properties (such as transition temperature, viscosity, etc.) and mechanical properties (such as strength, stiffness, and toughness) depend on the molecular weight of polymer. The lower the molecular weight, lower the transition temperature, viscosity, and the mechanical properties. Due to increased entanglement of chains with increased molecular weight, the polymer gets higher viscosity in molten state, which makes the processing of polymer difficult.

Weight-Average Molecular Weight

The weight-average probability is given by:

$$P_i = \frac{N_i M_i}{\sum_{j=o}^{\infty} N_j M_j}$$

The weight-average molecular weight is given by:

$$M_w = \sum_{i=o}^{\infty}\left[\frac{N_i M_i}{\sum_{j=o}^{\infty} N_j M_j}\right] M_i = \frac{\sum_{i=o}^{\infty} N_i M_i^2}{\sum_{j=o}^{\infty} N_j M_j}$$

A typical plot showing the number-average and weight-average molecular weight is shown in Figure. The number-average molecular weight is less than the weight-average molecular weight. The degree of polymerization can be calculated using the number-average molecular weight.

$$\text{Degree of polymerization} = \frac{\text{Number average molecular weight}}{\text{Molecular weight of the repeat unit}}$$

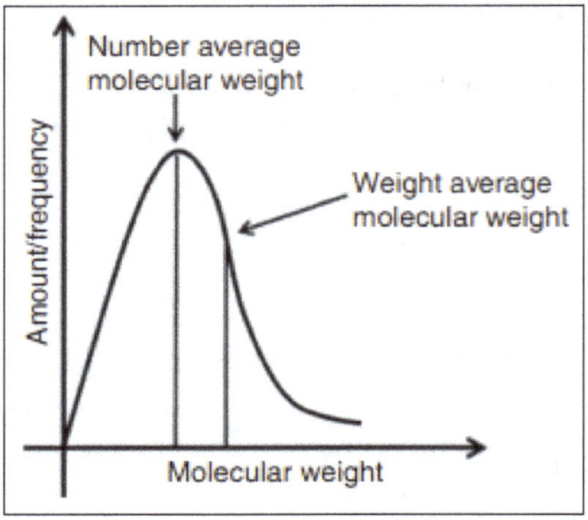

Figure: Average molecular weights of polymer.

Polydispersity Index or Heterogeneity Index

The ratio of the weight-average molecular weights to the number-average molecular weights is called polydispersity index (PDI) or heterogeneity index, which measures the polydispersity of the polymer mixture,

$$\text{PDI} = \frac{M_w}{M_n}$$

The dispersity measures heterogeneity of sizes of molecules or particles in the mixture. The mixture is called monodisperse if the molecules have the same size, shape, or mass. If the molecules in the mixture have an inconsistent size, shape and mass distribution, the mixture is called polydisperse.

The natural polymers are generally monodisperse as all synthetic polymers are polydisperse with some exceptions. The PDI is equal to or greater than 1 whereas the polymer chains approach uniform chain length, the PDI tends to unity.

Polymer Crystallinity: Crystalline and Amorphous Polymers

The polymeric chains being very large are found in the polymer in two forms as follows:

1. Lamellar crystalline form in which the chains fold and make lamellar structure arranged in the regular manner and amorphous form in which the chains are in the irregular manner. The lamellae are embedded in the amorphous part and can communicate with other lamellae via tie molecules. Polymer may be amorphous or semi-crystalline in nature. The % crystallinity is given by:

$$\%\text{Crystallinity} = \frac{\rho_c(\rho_s - \rho_a)}{\rho_s(\rho_c - \rho_a)} \times 100\ \%$$

- ρ_c = density of the completely crystalline polymer,
- ρ_a = density of the completely amorphous polymer,
- ρ_s = density of the sample.

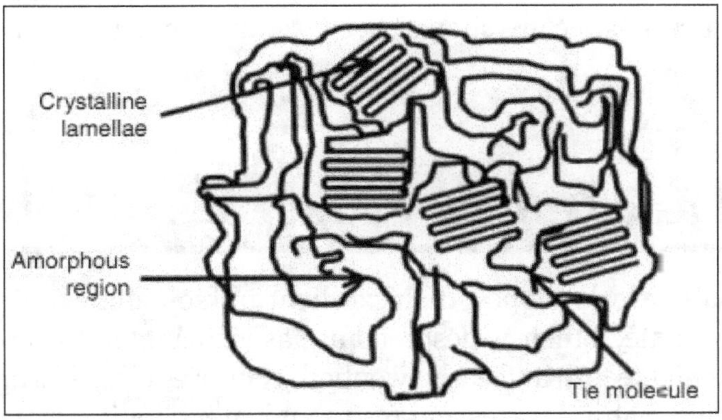

Figure: Semi-crystalline polymer.

2. A typical range of crystallinity can be defined as amorphous (0%) to highly crystalline (>90%). The polymers having simple structural chains as linear chains and slow cooling rate will result in good crystallinity as expected. In slow cooling, sufficient time is available for crystallization to take place. Polymers having high degree of crystallinity are rigid and have high melting point, but their impact resistance is low. However, amorphous polymers are soft and have lower melting points. For a solvent, it is important to state that it can penetrate the amorphous part more easily than the crystalline part.

- Examples of amorphous polymers: Polystyrene and poly(methyl methacrylate).
- Examples of crystalline polymers: Polyethylene, and PET polyester.

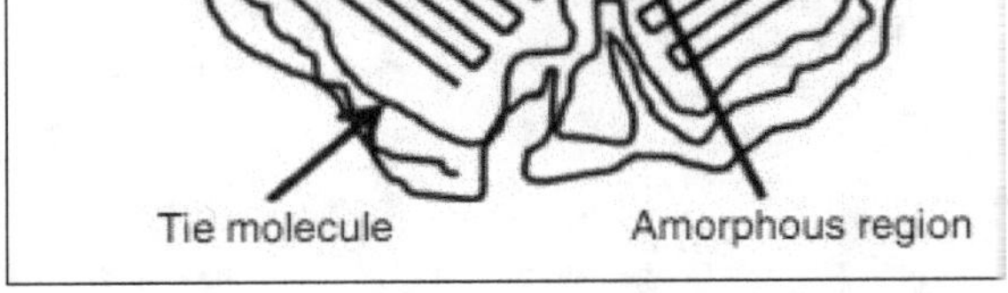

Figure: A typical structure of spherulite.

- Spherulites: If the molten polymer is cooled down, then the crystalline lamellae grow in radial direction from a nucleus along the three dimensions leading to a spherical structure called spherulite. The amorphous region is in between the crystalline lamellae. Spherulite formation and its diameter depend on various parameters such as the number of nucleation sites, polymer molecule structure and rate of cooling. Due to highly ordered lamellae in the spherulite, it shows higher density, hardness, tensile strength, and Young's modulus. The elasticity and impact resistance are shown, because the lamellae are connected to amorphous regions.

Electrical and Optical Properties of Polymers

As polymer materials have developed, their excellent and sometimes outstanding dielectric properties have guaranteed their widespread use as insulants in electrical and electronic engineering. In the nineteenth and early twentieth centuries electrical apparatus relied on wood, cotton sleeving, natural waxes and resins and later ebonite as insulating materials. Today a number of polymers including PTFE, PE, PVC, EP and MF offer an unrivalled combination of cost, ease of processing and electrical performance. These materials have played a most important part in the evolution of electrical components and equipment. Most electrical properties are determined largely by primary chemical structure, and are relatively insensitive to microstructure. In consequence the electrical behaviour of polymers is generally less varied than the mechanical behaviour. The same can be said of the optical properties, which nevertheless govern a variety of engineering end-uses.

Behaviour in a Steady (D.C.) Electric Field

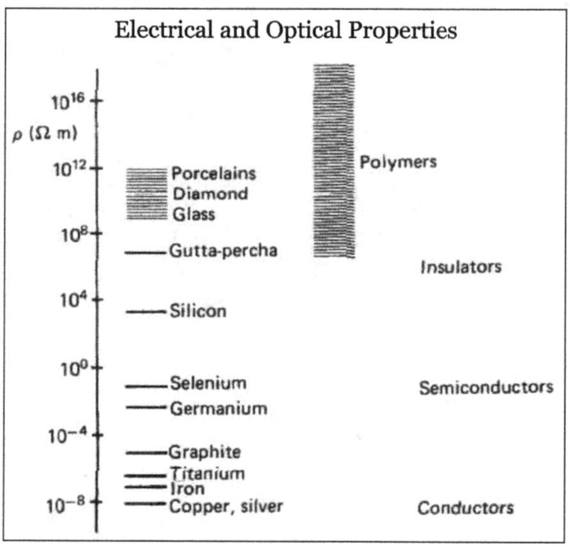

Figure: Volume resistivity ρ of electrical engineering materials.

The electrical properties of a material may be investigated by considering its response to imposed electric fields of various strengths and frequencies, just as the mechanical properties may be defined through the response to static and cyclic stress. We consider first the behaviour of polymers in steady (D.C.) electric fields.

Figure shows the range of volume resistivity p found in electrical engineering materials. Polymers as a class have the very high electrical resistivity characteristic of insulators. To quote resistivities in this way implies that Ohm's law is obeyed and that the conduction current $I = AE / \rho$, where E is the electric field strength and A the cross-sectional area of the material.

However, it is difficult to measure steady D.C. conduction in high performance insulants such as PTFE, where p may exceed 1016Ω m. The volume resistivity measured by standard test methods increases steadily with time. One-minute values of p are frequently quoted, but curves such as those of figure above showing how p changes with time of electrification present the d.c. conduction behaviour more satisfactorily. Conduction in the surface layers of a polymer material is often sensitive to ambient humidity and surface contamination. The surface resistivity is determined from the flow of current between two electrodes in contact with one surface of a thin specimen of polymer material.

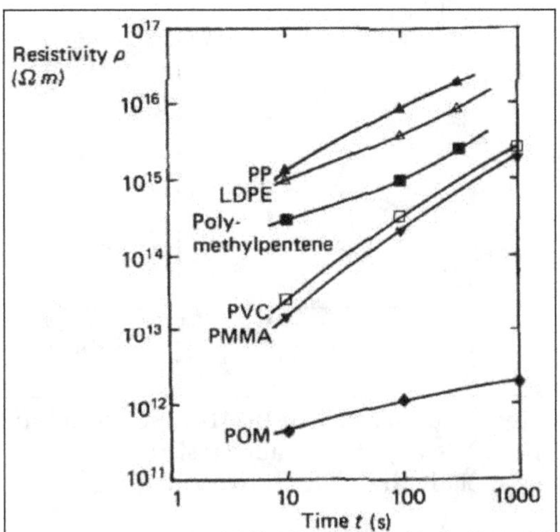

Figure: Changes in volume resistivity p with elapsed time for selected polymers.

The extremely low values of current at typical working voltages implied by the high values of p show the absence in polymers of any large number of charge carriers such as exist in metals and to a less extent in semiconductors. The valency electrons in polymer molecules (with a few exceptions) are localized in covalent bonds between pairs of atoms. The small currents which are observed to flow in weak fields arise from the movement of electrically charged species present as structural defects and impurities. The concentration of defects increases as the temperature rises, so that the resistivity falls. Exposure to ionizing radiation and absorption of water or plasticizer can also lead to an increase in the concentration of charge carriers with an accompanying increase in conductivity. The high conductivity of graphite demonstrates that polymeric materials are not invariably insulators. The conductivity of graphite (which is a form of pure carbon) arises directly from its chemical and electronic structure. In the solid the carbon atoms lie in parallel stacked layers. Within each layer the C atoms are bonded in hexagonal rings to form a continuous network of primary chemical bonds. However, a quarter of all the valence electrons in such a fused ring structure are not localized between pairs of atoms, but are delocalized, much as they are in metals, and are free to carry a conduction current. The conductivity of graphite is much higher parallel to the layers than perpendicular to them because the delocalization is essentially confined to the individual layers. Carbon fiber and other forms of polymeric carbon with unsaturated structures show similar high conductivity. In sharp contrast diamond, which is another form of pure carbon but which contains only localized electron pair bonds, is a dielectric. The hydrocarbon polymer polyacetylene has some electron delocalization along the chain and is a semiconductor with a resistivity of about 10^3 ohm m. It is likely that commercially important engineering polymers with intrinsic conductivity or semiconductor or photoconductor properties will be developed.

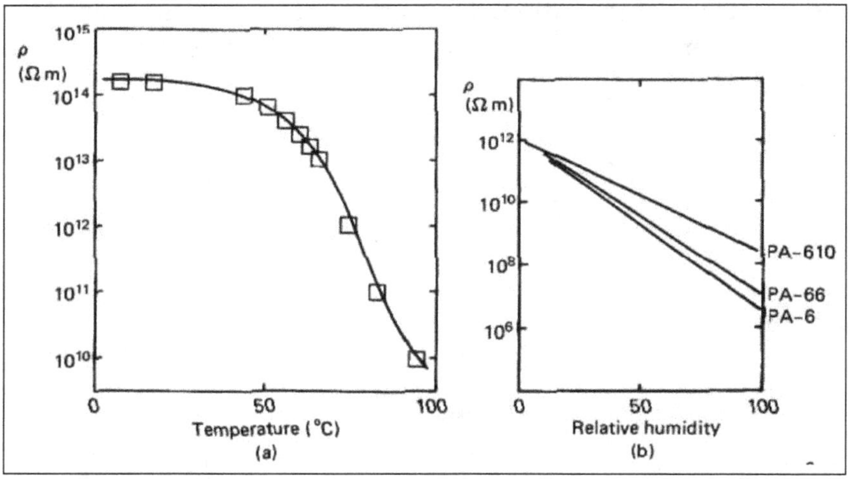

Figure: (a) Volume resistivity p of un-plasticised PVC between 0 and 100°C (all values measured after 60 s electrification), (b) Steady-state volume resistivity p of polyamides: dependence on relative humidity at room temperature.

In a perfect insulator no steady current flows in a static electric field, but energy is stored in the material as a result of dielectric polarization. The effect is analogous to the storage of mechanical energy in a perfect elastic material, and arises through the displacement of electric charge. Some polarization occurs in all materials through small displacements of electrons and nuclei within individual atoms, but larger effects arise if the solid contains permanent dipoles (from polar bonds or asymmetric groups of atoms) which tend to orient themselves in the direction of the externally imposed field. At normal electric field strengths the dielectric polarization is proportional to the field strength, and we are able to define an important linear property of the material the relative permittivity (or dielectric constant) $\varepsilon_r.\varepsilon_r$. Is the ratio e/eo of the permittivity of the material to the permittivity of a vacuum? The permittivity determines the size of the force acting between a pair of electric charges separated by the dielectric material. The capacitance C of a parallel plate capacitor is proportional to the relative permittivity of the medium between the plates. The energy stored by a capacitor charged at voltage V is ½ CV2 and therefore the energy stored by a dielectric in an electric field increases with its permittivity.

The D.C. relative permittivities of selected polymers are listed in table below. The values are determined largely by the nature and arrangement of the bonds in the primary structure. In polymers such as PE, PP and PTFE there is no dipole in the mer because of symmetry. Furthermore both bonding and non-bonding electrons are tightly held and displaced little by external fields. These materials exhibit very little dielectric polarization, and as a result e_r are low. Polar polymers such as PMMA, PVC and notably PVDF possess higher values of e_r.

The insulating property of any dielectric breaks down in sufficiently strong electric fields. However, in polymers the dielectric (or electric) strength may be as high as 1000 MV/m. An upper limit on dielectric strength is set by the ionization energies of electrons in covalent bonds within the polymer primary structure. Purely electrical or intrinsic breakdown occurs when appreciable numbers of electrons are detached from their parent molecules and accelerate in the electric field to cause secondary ionization and avalanching. Breakdown of this kind runs its course extremely rapidly and the breakdown voltage does not depend greatly on temperature.

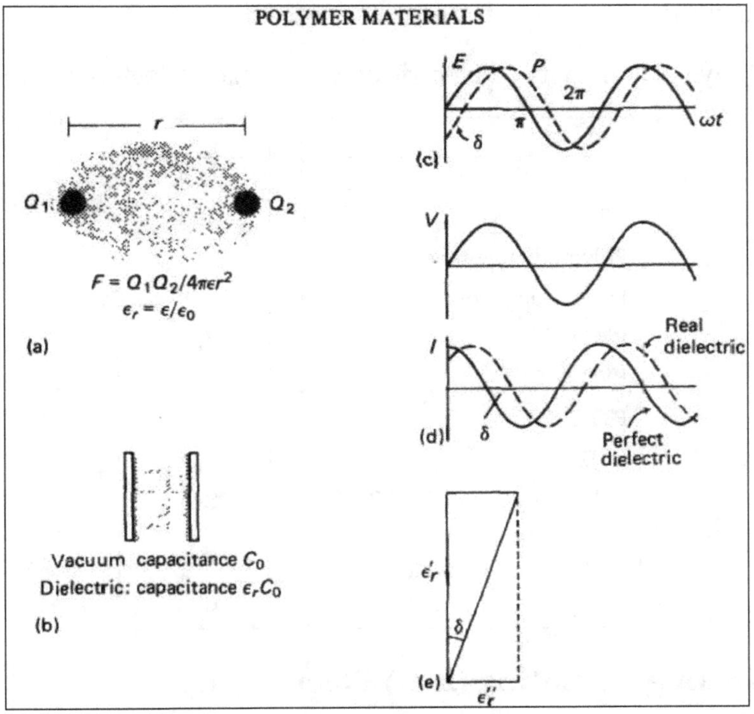

Figure: Relative permittivity ε_r.

(a) Screening of electric charges Q1 and Q2 by a dielectric of permittivity ε. (b) Capacitance of a parallel plate condenser with dielectric, (c)-(e) Response of a dielectric in an alternating electric field, (c) Polarization response P in a sinusoidal electric field B. (d) Voltage—current relation of a perfect (lossless) and of a real dielectric, showing phase angle 6. (e) Real and imaginary parts of the complex relative permittivity e_r^*.

Dielectric breakdown may occur at lower electric field strengths for several reasons. If power dissipated in the dielectric is not lost to the surroundings the rising temperature may bring about thermal breakdown. Breakdown voltage in this case depends on heat loss, and hence on the geometry of the specimen and the ambient temperature. In some materials, such as PE above 50 °C, the polymer becomes severely compressed in intense electric fields, and failure may occur through mechanical collapse. Surface contamination (for example by dust and moisture) may lead to local breakdown by tracking, a mode of failure in which carbon conducting paths are formed across the surface by localized pyrolysis of polymer. Polymer materials differ greatly in their tendency to fail by tracking, PTFE and PE/PP elastomer (EPM) having particularly good resistance to tracking. Alternatively, discharge through air or vacuum from an electrode to the polymer surface may cause erosion of the material; if the solid polymer contains voids (as in the case of sintered granular PTFE) such discharge erosion may penetrate through the material, and reduce the dielectric strength considerably. Standard tests for dielectric strength do exist, but measured breakdown voltages should be regarded as markedly dependent on test and specimen conditions.

Table: D.C. relative permittivity (dielectric contestant) of selected polymers.

	ε_r
PE	2.3
PP	2.3
Polymethylpentene	2.1
POM copolymer	3.8
PMMA	3.8
PVC	3.8
PTFE	2.1
EPDM	3.1
Chlorosulphonated PE (CSM elastomer)	8-10
Urethane elastomer (AU, EUo	9
PVDF	9-13

Behaviour in an Alternating (a.c.) Electric Field

We defined a complex compliance D* = D' - iD", and showed that D"/D' = tan δ, the loss tangent. The dynamic mechanical behaviour of the material is described by the frequency dependence of D*.

In a similar way the linear response of a dielectric to an alternating electric field is described by the use of a complex relative permittivity $\varepsilon_r^* = \varepsilon_r' - i\varepsilon_r''$. The ratio e"/e', is the dielectric loss tangent tan 5 (known also as the dissipation factor). 8 is the phase lag between the electric field and the polarization of the dielectric. In a perfect dielectric (which behaves as a pure capacitance) the phase angle between current and voltage is $\pi/2$. In real dielectrics, δ is not zero, and current leads voltage by $(\pi/2 \sim \delta)$. Sinδ is known as the power factor. Finally we note that $\varepsilon'' = \varepsilon_r' \tan\delta$ is the dielectric loss index of the material.

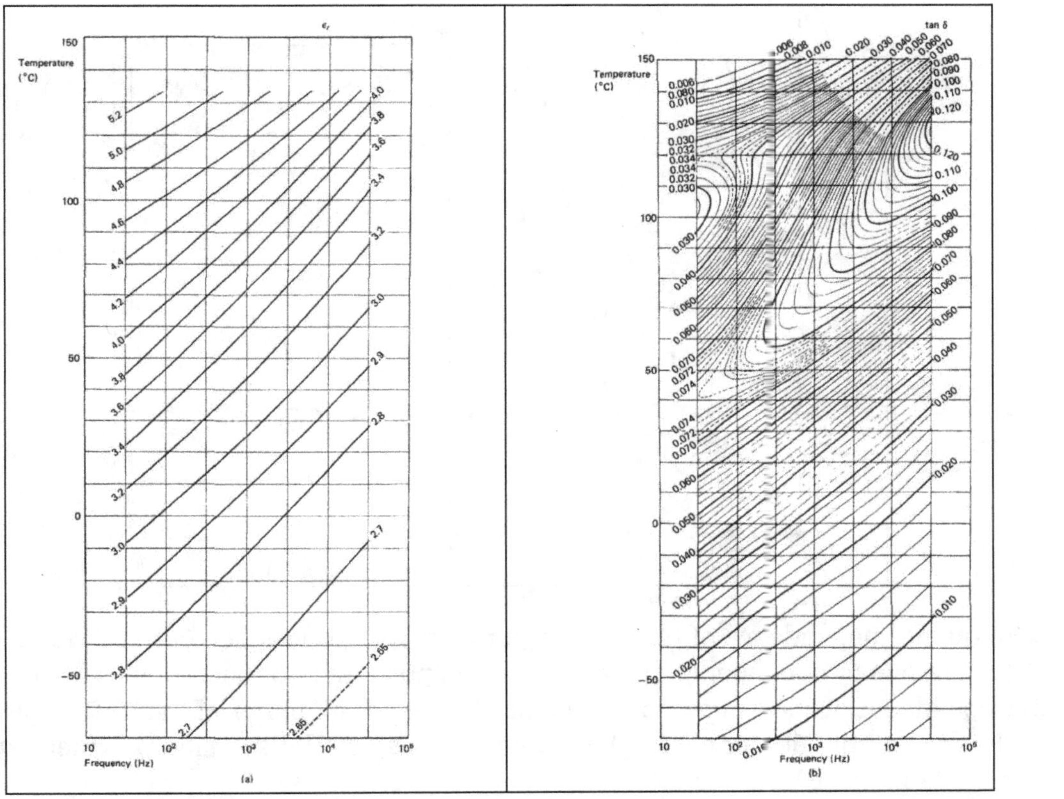

Figure: Frequency and temperature dependence of (a) permittivity and (b) loss tangent of PMMA (cast sheet, dry).

The a.c. electrical properties of polymers are thus commonly and conveniently expressed in the form of relative permittivity ε_r' and loss tangent $\varepsilon_r''/\varepsilon_r'$, data as functions of frequency (and temperature). Data may extend from d.c. or low audio frequency (30 Hz) in half-decade steps to high radiofrequency (1000 MHz). In pure homogeneous nonpolar polymers, the polarization arises almost entirely from displacements of electrons and nuclei. These redistributions occur extremely rapidly, in picoseconds or less, and the polarization response follows the alternating electric field without lag up to high radiofrequencies and beyond. As a result polymers of this type, notably PTFE and PE, show little frequency dependence of ε_r' and extremely low loss tangents with no loss peaks. In homogeneous PTFE tan 8 may be as small as 10-5 and since $\tan\delta \approx \delta$ for small δ the loss tangent (or loss angle) is often expressed in microradians. The loss tangents of pure PE and PTFE are so low that they may be sharply increased by very small concentrations of impurities and additives, or by physical heterogeneity. PEs synthesized by different processes may show different loss characteristics depending on the number of CO impurities incorporated in the chain. Sintered granular PTFE shows a loss peak which is absent in material of high crystallinity. PTFE with high filler content may have a loss tangent as high as 0.1 in damp conditions.

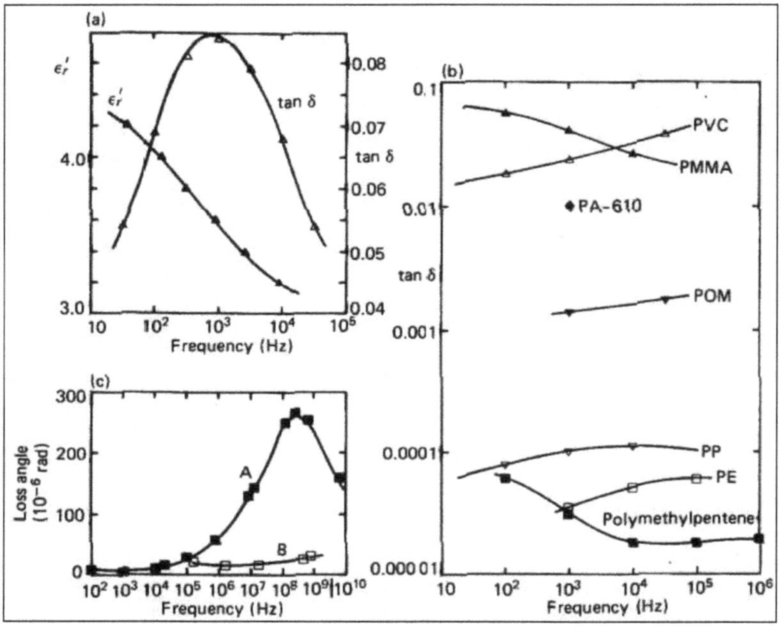

a) Frequency dependence of the relative permittivity and loss tangent of PMMA at 70 °C. (b) Frequency dependence of loss tangents of polar and nonpolar polymers at 20 °C. (c) Dependence of loss tangent on frequency for two PTFE materials: sintered granular PTFE, crystallinity 60 per cent (A); unsintered coagulated dispersion polymer, 93 per cent crystallinity (B).

The loss tangents of polar polymer materials are generally much higher than those of nonpolar polymers. As figure above shows, loss peaks are observed at frequencies corresponding to the relaxation times for dipole reorientation and the positions of these loss peaks can often be correlated with known structural transitions, such as T_g and T_m. The permittivity and loss in the vicinity of a loss peak may be markedly affected by changes in temperature.

The extremely low dielectric loss of PE and PTFE make these outstanding materials for high frequency electronic devices and components. The lossy polymers such as PVC are valuable insulants (for example as cable sheathing) in low frequency applications. At higher frequencies dielectric heating (and the possibility at high voltages of thermal breakdown) may become serious in high loss polymers.

A heated polymer polarized in a very strong electric field and then cooled may retain its polarization indefinitely. Such a polarized dielectric is the electrostatic counterpart of a magnet, and is known as an electret. Polymeric electrets can be prepared with field strengths of 30 kV/cm, showing little loss of strength over periods of several years. Electrets bearing a net electric charge stable for long periods can be made from suitable polymer materials carrying metal foil on one face. Such foil-electrets are used, for example, in microphones.

Finally, we mention the static electrification or triboelectric charging of polymers by contact or friction. The high resistivity of polymers allows large electrostatic charges to accumulate. This has many troublesome consequences (as in textile manufacture and use) and is occasionally hazardous. The study of how triboelectric charging depends on polymer structure is in its infancy, but table below shows a triboelectric series which indicates the sign of the electrostatic charge acquired by contact between pairs of dissimilar polymers. Antistatic formulations of polymers have much reduced resistivities so that electrostatic charge may be dissipated by leakage currents.

Table: Triboelectric series.

Positive end	Wool PA Cellulose Cotton Silk CA
Negative end	PMMA PVAL PETP PAN PVC PVDC PE PTFE

Optical Properties

The highest a.c. electrical frequencies we have discussed fall within the short radio wave and microwave regions of the electromagnetic spectrum. The transmission and absorption of radiation by the polymer dielectric are determined by the quantities e, and e". Even higher frequencies bring us into the infrared and visible parts of the spectrum, that is, to optical phenomena.

There is an important underlying continuity in the electrical and optical behaviour of materials; for example, the optical refractive index n and the high frequency permittivity are linked in electromagnetic theory by the relation $n = (\varepsilon_r^i)^{1/2}$. At optical frequencies as at lower a.c. frequencies absorption of radiation occurs by irreversible non-radiative loss processes. As we shall see, absorption of optical radiation occurs by excitation of bond vibrations (infrared) and rearrangements of electrons within molecules (visible and ultraviolet). Even so we should not overstress the internal connections between electrical and optical properties, which in practice are usually considered separately and have distinct terminologies. In addition we shall see that scattering phenomena play a major part in determining optical properties. Light scattering is a manifestation of optical diffraction processes, which occur prominently in polymer materials having microstructural features of dimensions comparable with optical wavelengths (0.5 / μm). Diffraction effects within materials are rarely important at electrical a.c. frequencies because of the much greater associated wavelengths.

Color and Infrared Absorption

Very few pure polymers absorb radiation in the visible spectrum, that is, roughly between 380 and 760 nm. Thus most polymers are colorless. The only exceptions are some thermosets and elastomers, including PF, some polyurethanes, epoxies and furan resins, which absorb more or less strongly at the blue end of the spectrum and consequently appear brownish when viewed by transmitted or reflected light. These substances contain alternating double and single covalent bonds or aromatic rings which act as chromophores, absorbing light at frequencies corresponding to the excitation energies of bonding electrons. Graphite and the other polymeric carbons which have fused ring structures and which appear intensely black exhibit this absorption throughout the visible spectrum in extreme form. Deliberately colored polymer materials are invariably produced by incorporating as additives either finely divided colored solids (pigments) or soluble colored substances (dyes). The colors of solid polymers are fully defined by their spectra, measured either by transmission or reflection as appropriate. It is often more useful for purposes of specification to refer to colors by using the CIE trichromatic system of color measurement.

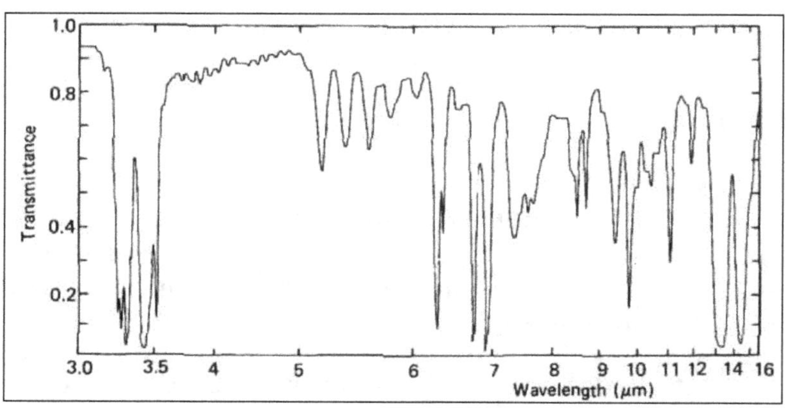

Figure; Infrared spectrum of polystyrene measured in transmission.

Although many polymers show no absorption of radiation at visible wavelengths, they invariably absorb strongly at certain frequencies in the infrared. The infrared (transmission or reflectance) spectrum of a polymer is determined by its molecular structure. Infrared spectroscopy is one of the most powerful methods of polymer analysis, enabling polymer materials of unknown composition to be identified very rapidly from small samples. Figure above shows the infrared spectrum of polystyrene PS. The sharply defined absorption peaks correspond to various modes of vibration of individual chemical bonds, and give the analytical chemist detailed information about the primary molecular structure. The infrared spectrum is a fingerprint which allows even closely similar materials such as the various polyamides to be distinguished.

Refraction

The visual appearance and optical performance of a polymer material depends, apart from color, on the nature of its surface and its light transmission properties.

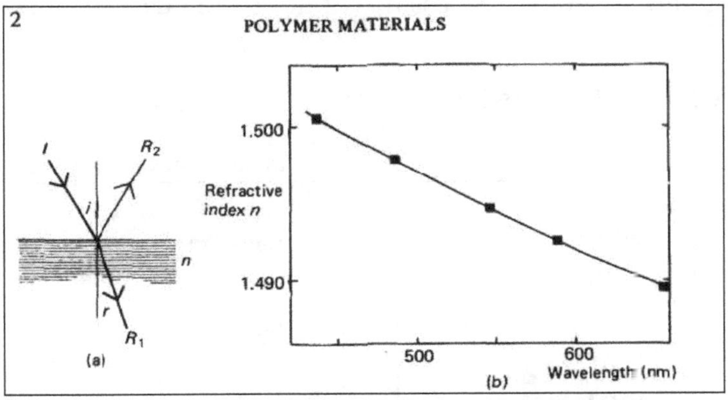

Figure: (a) Refraction and reflection at the surface of a transparent solid.
(b) Variation with wavelength of the refractive index n of PMMA.

The optical phenomena which occur when a ray of light strikes the plane surface of a colorless transparent material, are governed by the refractive index n. Snell's law relates the angle r to the angle i; Fresnel's equation allows the intensities and polarizations of the refracted and reflected rays R_1 and R_2 to be calculated from n and i. n varies with wavelength and accurate determinations are made with monochromatic light at specified wavelengths, using selected lines from the atomic emission spectra of elements contained in discharge lamps. The wavelength dependence of the refractive index, which leads to refractive dispersion of white light, is expressed in terms of the reciprocal dispersive power V_d of the material:

$$V_d = \frac{n(d)-1}{n(F)-n(C)}$$

Where d denotes the helium emission line at 587.6 nm; and F and C denote hydrogen lines at 486.1 and 656.3 nm.

Table below lists the refractive indices of a number of polymers. More precise data for polymers which find use as optical materials are given in table which includes for comparison two optical glasses.

Refractive index is determined by the extent to which the electronic structure of the polymer molecules is deformed by the optical frequency electric field of the radiation. If a material is structurally isotropic, as in the case of unstressed amorphous polymers, then it is also optically isotropic, and a single refractive index characterizes the refraction behaviour. In crystals and other anisotropic materials the refractive index takes different values along different principal axes, and the material is said to be doubly refracting or birefringent. Amorphous materials under deformation develop birefringence as molecules become aligned. Similar effects arise in flowing polymer melts. The study of birefringence is thus a very useful method of exploring the microstructural effects of deformation. One particularly important technical application of stress-induced birefringence in polymer materials is in photoelastic stress analysis, a method of investigating complex stress distributions in engineering components and structures.

Table: Refractive index of selected polymers.

	Refractive index
PTFE	1.35
PCTFE	1.42-1.43
PVDF	1.42
CA	1.48-1.50
POM	1.48
PMMA	1.49
PP	1.49
PF	1.5-1.7
EP	1.5-1.6
LDPE	1.51
PAN	1.52
NR	1.52
PA	1.53-1.55
HDPE	1.54
PVC	1.54-1.55
UF	1.54-1.56
CR	1.55
PS	1.59
PC	1.59
PVDC	1.60-1.63

Table: Refractive inde4x n(d) and reciprocal dispersive power V4 of optical materials.

	N(d)	Vd
PS	1.591	31.0
PC	1.586	29.9
PMMA	1.492 6	59.3
Allyl diglycol carbonate	1.498	53.6
Polymethylpentene	1.465 4	44.9
SAN	1.569	35.7
Borosilicate crown glass	1.509 7	64.4
Dense flint glass	1.620 5	36.2

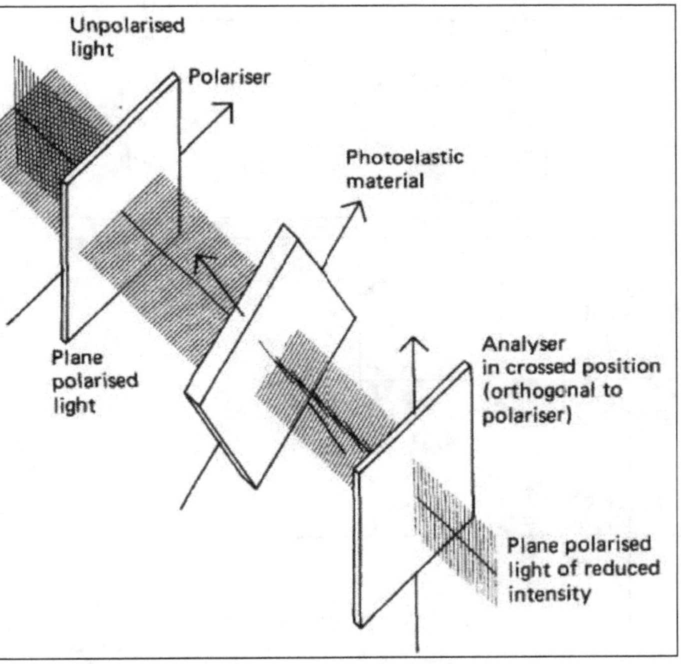

Figure: Principle of photoelastic stress analysis.

Figure above summarizes the principle of a simple photoelastic analysis. The component under examination is modeled in a stress-free polymer material either by molding or machining. The model component is then loaded and viewed by transmitted light between crossed Polaroids. Birefringence induced by the internal stress field generates two components of the refracted ray which travel at different speeds, and may thus produce interference effects. For example, variations in stress within the component give rise to a fringe pattern from which the principal stress lines can be obtained. The value of photoelastic analysis depends almost entirely on the availability of polymeric materials (notably epoxies, polyesters, PC, PS and PMMA) with a suitable combination of mechanical and optical properties. The change AM in refractive index produced by a stress P defines the stress-optical coefficient C of the material by the relation $\Delta n = CP$.

Transparency

Colorless polymeric materials range from highly transparent to opaque. Loss of transparency arises from light scattering processes within the material, which distort and attenuate the transmitted image. Figure below shows a light ray I striking a plane surface at right angles. A small fraction of the incident flux is reflected back from the surface at the point of incidence p. The ratio of reflected to transmitted flux is calculated by Fresnel's equation and for normal incidence $\phi(R_1)/\phi(T)=(n_2-n_1)^2/4n_1n_2$. at an air-polymer interface, with n_2 = 1.5, about 4 per cent of the incident flux is reflected. If the air is replaced by a transparent liquid having the same refractive index as the polymer the surface reflection at p (and also at r) is eliminated.

	10^{12} C(m^2 N^{-1})
Selected EP resins	55
PMMA	-4.7
PC	78
Polyurethane elastomer	3500

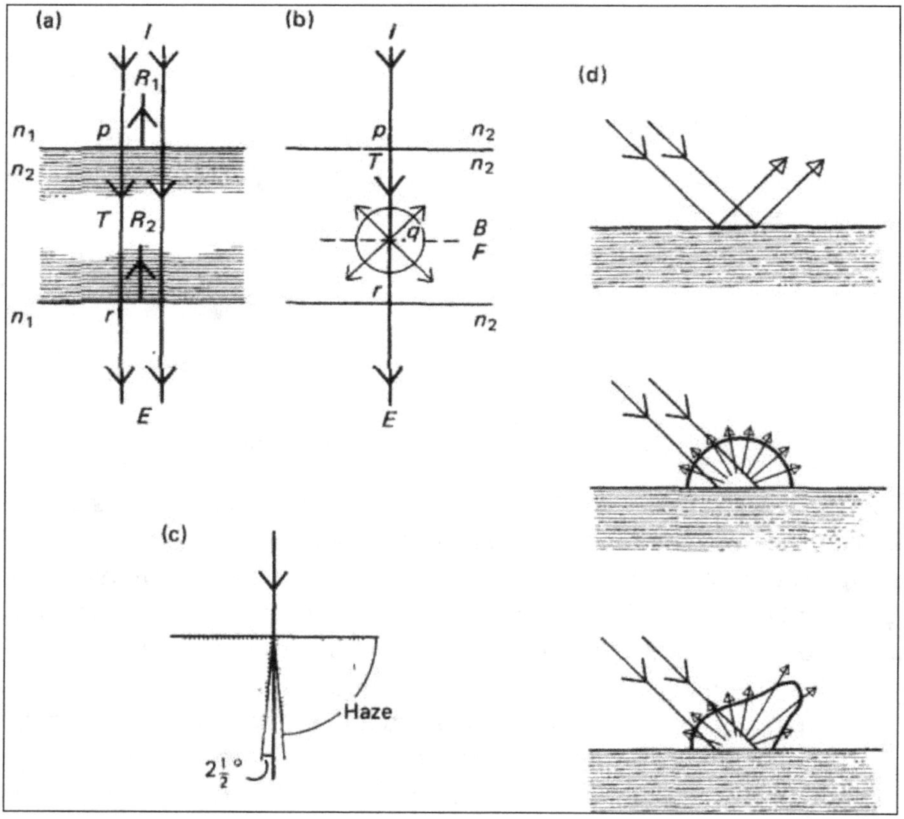

Figure shows (a) Reflection and (b) Scattering of light on transmission through a transparent medium, (c) Haze produced by wide-angle forward scattering of light, (d) Gloss: (from the top) specular reflection; perfect diffuse reflection; intermediate behaviour. The polar envelope shows the angular distribution of reflected light intensity.

If the material is non-absorbing and optically homogeneous and isotropic the transmitted ray propagates without loss of intensity. In real materials the transmitted ray may lose some intensity as it travels forward as the result of scattering from refractive index inhomogeneities within the material. Such scattering is akin to the ray splitting which occurs at the surface at p and which is repeated whenever the ray encounters a change of refractive index δn_2 at point's q within the material. The angular distribution of the scattered light is unlikely to be simple, but it is useful to speak of back-scattered and forward scattered components B and F of the total flux. We have $\phi(I) = \phi(B) + \phi(F) + \phi(E) < p$, where E is the un-deviated emergent ray. The direct transmission factor T of the material is determined as the ratio $\phi(E)/\phi(I)$. For weakly scattering colorless materials T decreases exponentially with the sample thickness /, and therefore inT = $-\sigma l$, where a is the scattering coefficient.

The effect of scattering is to reduce the contrast between light and dark parts of the object viewed through the transparent polymer, and thus to produce haze or milkiness in the transmitted image. The direct transmission factor, the total transmitted flux ϕ_τ, i.e. the sum of $\phi(F)$ and $\phi(E)$, and the forward scattered fraction $\phi(F)/\phi_\tau$ can be determined photometrically.

Besides haze, scattering may produce loss of clarity so that distinct features of the object can no longer be distinguished in the transmitted image. Loss of clarity is caused by forward scattering at angles very close to the un-deviated beam, since it arises from confusion of light rays diverging from the object at the limits of the angular resolution of the eye. Haze on the other hand is associated with scattering over a wide range of angles. A quantitative measure of haze is obtained in an ASTM standard test from the total forward-scattered flux, excluding that within 2½ degrees of the un-deviated ray.

In polymer materials heterogeneity of refractive index can arise from differences of density in amorphous and crystalline regions within the polymer itself, or from solid particles incorporated as pigments or fillers, or from voids. The degree of scattering depends strongly on the variation of refractive index ($\pm \delta n_2$) and on the size of the heterogeneities. The most efficient scattering occurs when the scattering centers are comparable in dimensions with the wavelength of light. The scattering of light in pigmented polymer films has been extensively studied in connection with the optical properties of paints. Crystalline polymers are normally translucent or opaque unless it happens (as in polymethylpentene) that there is little difference in the refractive indices of crystalline and amorphous regions or (as in cellulose triacetate) that the spherulites are exceptionally small. It is frequently possible to enhance the transparency of polymer materials by assisted nucleation or by rapid cooling from the melt, both means of decreasing the spherulite size. Stretching is also an effective way of increasing transparency, since spherulites are transformed into oriented fibrils which scatter light less effectively.

Table: Selected optical properties of some engineering polymers.

	PP	LDPE (rapidly cooled film)	Polymethylpentene (modified)	PA (nylon 66)	PMMA	PVC	PTFE
Refractive index n (546.1 nm)	1.505	1.516	1.472	-	1.495	-	-
Refractive index n (589.3 nm)	-	-	1.465	1.448	-	1.542	1.325
Temperature coefficient of refractive index (10^4 dn/dT)Fk^{-1})	-	-2.4	-2.4	-	-1.0	-1.1	-
Direct transmission factor T%) (normalized to 1 mm sample thickness)	11	45	99	0	99	94	0
Scattering coefficient 10^{-2} $\sigma(m^{-1})$	21.5	8.3	0.11	100	0.01	0.6	80

Gloss

'Gloss' is a term applied to the surface optical character of a material, whether transparent or opaque. A perfect mirror-like surface, a specular reflector, shows one extreme of behaviour. At the other extreme, a highly scattering surface (a perfect diffuse reflector) reflects light equally in all directions at all angles of incidence.

The gloss of real surfaces is most completely described by the angular intensity distribution of reflected light for one or more angles of incidence. A somewhat simpler index of surface appearance is the direct reflection factor, the ratio of the flux reflected at the specular angle to the incident flux, for angles of incidence from 0 to 90 degrees. Table above collates optical data on a number of engineering polymers.

Light Transfer

Strongly scattering translucent materials (optical 'diffusers') are unable to transmit useful images but nevertheless may transmit appreciable amounts of diffuse light. Their optical characteristics are most simply expressed through the total transmission factor 7" (or luminous transmittance) for normal incidence, which is defined as $[\phi(F) + \phi(E)]/\phi(I)$.

References

- Effect-of-molecular-weight-on-polymer-properties: polymerdatabase.com, Retrieved 26, April 2020
- Mechanical-properties-of-polymers: acs.org, Retrieved 03, June 2020
- Physical-thermal-and-mechanical-properties-of-polymers: wiley.com, Retrieved 19, January 2020
- Electrical-and-Optical-Properties: homepages.ed.ac.uk, Retrieved 22, August 2020

3

Organic Polymers

Organic polymers are the polymers which contain carbon atoms in the backbone. Organic polymers play a crucial role in living organisms like providing basic structural materials and participating in vital life processes. This chapter has been carefully written to provide an easy understanding of the varied facets of organic polymers.

Organic polymers are materials that essentially contain carbon atoms in the backbone. Organic polymers have carbon-carbon bonds along with hetero-chain of atoms such as oxygen, nitrogen and sulphur inserted along the backbone. In most of the aqueous solutions, organic Polymers are typically poor conductors of electricity and heat. There are greater intermolecular forces between the long chains of organic polymers and the melting and boiling points of organic polymers is high. Organic polymers have carbon-carbon bonds between molecules and hence are insoluble in water, though they are soluble in other organic solvents. Organic polymers have wide variety of uses, for example: polystyrene resins are used in the production of home electronics and appliances; nylon-6 is used in textile and plastic industries. Organic polymers such as polyethylene terephthalate are in the manufacture of popular PET bottles. Others such as neoprene are used in shoe soles and wet suits, polyvinyl chloride in pipes and Teflon in non-stick pans.

Organic polymers can be synthesized by various polymerization procedures. The synthesis by bulk polymerization involves a homogeneous system where the reagents, mixed in the reaction medium, polymerize via radicular reactions. The synthesis must be made in inert atmosphere, since the presence of oxygen can delay the polymerization process. The polymerization reaction is induced by heating or UV radiation, depending on the employed monomer in the synthesis. Several monomers can be used, in accordance to the desired application and interaction with the analyte, as shown in Table. Bulk polymerization produces a monolithic rigid polymer matrix, needing to be broken into large, nonuniform particles in which the binding sites are distributed heterogeneously in the polymer matrix. Usually, the product obtained is dried, macerated and, depending on the use, separated by particle size. Generally organic polymers are flexible, presenting low density and stability over a wide pH range. Packed in columns or cartridges, can provide resistance to passage of liquids (occurring swelling). To avoid rigid matrix formation and the grinding step, it is possible to adopt other synthesis procedures such as polymerization by suspension or by precipitation. The procedures adopted in these other techniques contemplate reactions in a heterogeneous environment and allow obtaining particles with uniform sizes.

Table: Monomers Typically Used in the Preparation of Organic Polymers with Functional Matrix and Their Possible Interactions with the Analyte.

Functional Monomers	Monomers Name	Type of Interaction with the Analyte
	Acrylic acid	Ionic interaction and hydrogen bonding.
	Methacrylic acid	Ionic interaction and hydrogen bonding.
	p-vinylbenzoic acid	Ionic interaction and hydrogen bonding.
	Acrylamidosulfonic acid	Ionic interaction.
	4-Vinylpiridine	Ionic interaction, hydrogen bonding and charge transfer.
	2-Vinylpiridine	Ionic interaction, hydrogen bonding and charge transfer.
	4-Vinylimidazole	Ionic interaction, hydrogen bonding and coordination with metals.
	1-Vinylimidazole	Ionic interaction, hydrogen bonding and coordination with metal.
	Acrylamide	Hydrogen bonding.

Performing the polymerization procedure by suspension, the polymerization reactions take place inside the solvent droplets (immiscible solvent) where the monomer is dispersed. This procedure, however, must be performed in two steps, with the first being the solubilization of the monomer, cross-linking reagent, and radical initiator in an organic of polar solvent. Subsequently, the second is to perform the addition of this reaction mixture in water with a stabilizing agent to avoid the monomer drops coagulation during the process. Thus, the reaction occurs under constant heating and stirring, resulting in particles containing uniform dimensions. In the polymerization method by precipitation, the procedure is similar to the adopted in the bulk polymerization reaction, with the main difference in the solvent amount used. Thus, a solvent volume five times higher is used, preventing the monolith formation. In this method, the polymerization occurs with the growth of the polymer chain, making the polymer more insoluble in the reaction medium, forming a suspension with milky aspect at the reaction end, constituted by particles of micrometric order.

Performing the synthesis, the solvent type must be considered due its interactions with the reagents. If the reactive monomers present high solubility in the polymer chain and lower solubility in the porogenic solvent, the formation of spherical polymeric particles containing pores occurs after the solvent evaporation. If there is a high solubility of the monomers into the porogenic solvent, a more compact polymer matrix will occur, with a smaller number of pores and probably with a smaller surface area. During the polymerization reaction, the radical initiator breaks its structure by heating or by radiation, forming radicals in which are responsible by the polymerization reaction beginning. There are several radical initiators to be used in the polymerization reactions as: 2,2′-azo-bis-isobutyronitrile (AIBN); 2,2′-azo-bis(2,4-dimethylvaleronitrile) (ABDV); benzyl dimethyl acetal; benzoyl peroxide (BPO); and 4,4′-azo-bis (4-cyanopentanoic acid). The cross-linking reagent is responsible to the polymer thermal and mechanical stability, ensuring the rigidity of the bonding sites. The presence of the cross-links in the polymer matrix can be adjusted by varying the amount of the cross-linker in the reaction medium. The rigidity and stability achieved by crosslinking allows the use of the polymer in various analytical media without loss of its properties and avoiding the material degradation. Examples of cross-linking reagents are: *N*, *N'*-1,4-phenylenediacrylamine, ethylene glycol dimethacrylate, divinylbenzene, trimethylpropane trimethacrylate, trimethylolpropane trimethacrylate, *N,N'*-methylenediacrylamide, tetramethylene dimethacrylate, and among others.

Vinyl Polymer

Vinyl polymers are polymers made from *vinyl monomers*; that is, small molecules containing carbon-carbon double bonds. They make up largest family of polymers. Polyethylene is made from the monomer *ethylene*, which is also called ethene. When polymerized, the ethylene molecules are joined along the axes of their double bonds, to form a long chain of many thousands of carbon atoms containing only single bonds between atoms.

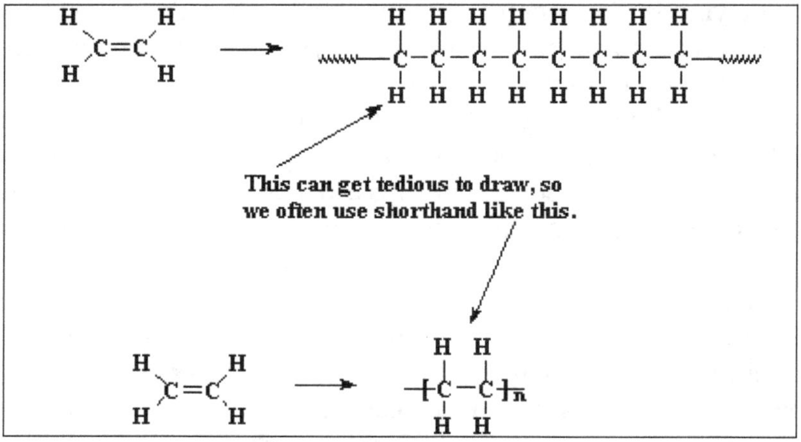

More sophisticated vinyl polymers are made from monomers in which one or more of the hydrogen atoms of ethylene has been replaced by another atom or groups of atoms.

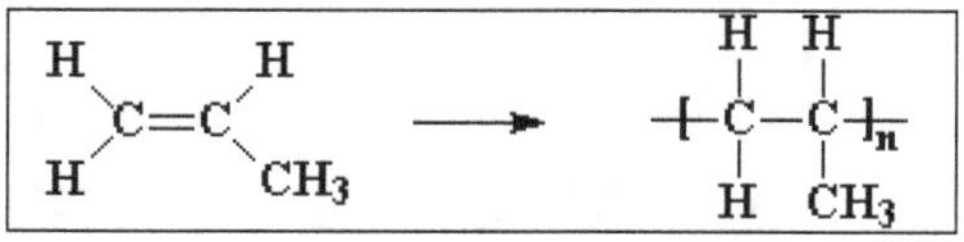

Figure: Polypropylene

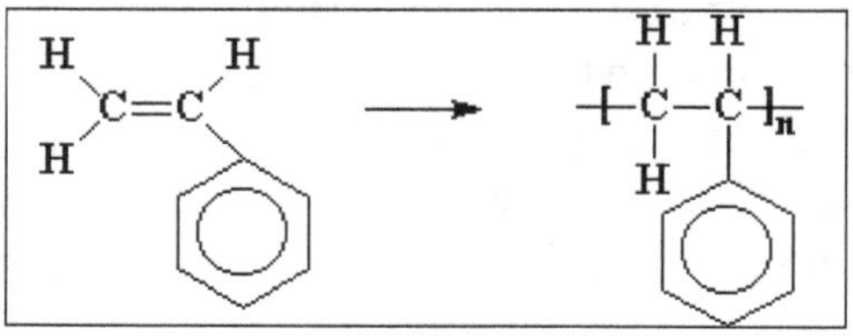

Figure: Polystyrene

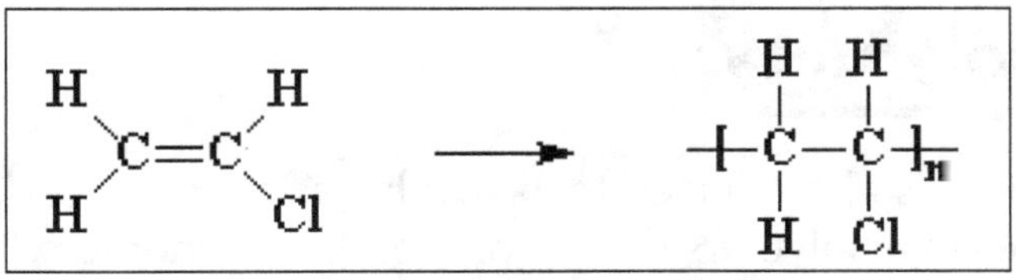

Figure: Poly(vinyl chloride)

Replacing two hydrogen atoms, on the same carbon atom, we can get polyisobutylene, which is type of rubber.

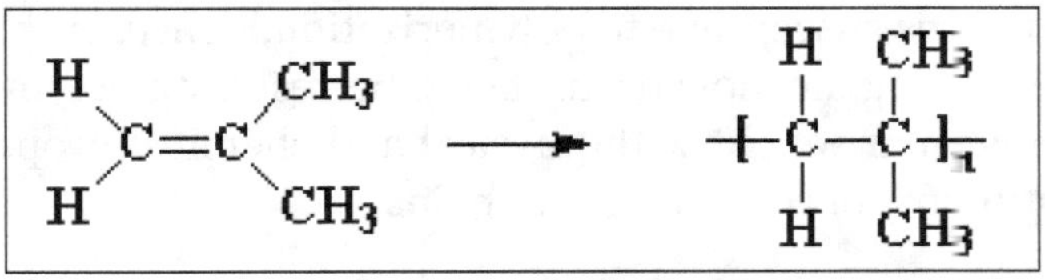

Figure: Poly(methyl methacrylate)

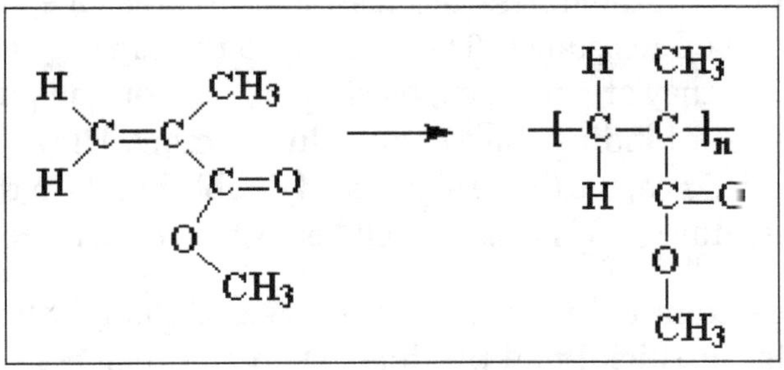

Not many monomers in which hydrogen atoms have been replaced on both carbon atoms will polymerize. But one polymer that is made from a monomer substituted on both carbon atoms is polytetrafluoroethylene, which DuPont makes and calls Teflon.

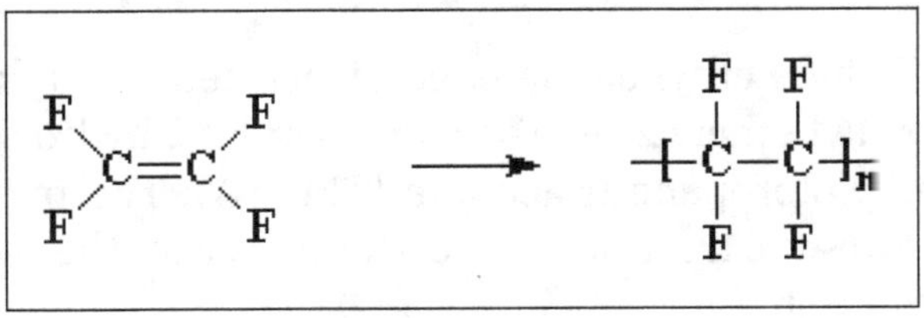

Vinyl polymers are made from vinyl monomers in a variety of ways, like:

- Free radical vinyl polymerization.
- Anionic vinyl polymerization.
- Cationic vinyl polymerization.
- Ziegler-Natta catalysis.
- Metallocene catalysis polymerization.

Acrylate Polymers

Acrylic polymers can also be referred to as acrylates. Acrylates (or polyacrylates), belong to the family of vinyl polymers made from acrylate monomers. They are the esters, salts and conjugate bases of acrylic acid and its derivatives, of which methacrylates are the most common members. Its monomers are esters containing vinyl groups, which comprises two carbon atoms that are double-bonded to each other and directly attached to the carbonyl carbon of the ester group. This makes acrylates to be bifunctional in nature: the vinyl group is susceptible to polymerization, hence; carbonyl group carries myriad functionality using alcohols or amines. The resulting polymers are characterized with a high molecular weight with physical and chemical properties that depend on the lateral substituents of the polymeric chains.

Acrylates are important class of polymers that are rubbery, soft and tough. They are well-known for their good impact toughness and resistance to breakage, transparency, elasticity, fairly good heat and oil resistance. They have also the capacity to endure a temperature of 170–180°C when they are in oil, exposed to dry heat or non-surfactant suspending agents that have anti-static binding abilities and film-forming. They also have good weatherability and ozone resistance due to the absence of double bonds in their backbone. Their glass transition temperature (T_g) is usually well below the room temperature.

Acrylate polymers are commonly used in dentistry and many other biomedical applications, cosmetics and artificial nail products such as eyelashes, nail enhancing polishes, nail builders, artificial nails and to help artificial nails mold to the natural nail plate as an adhesive, hair fixatives, in marine phytoplankton as a poisonous defense against predators such as protozoa, and oil seals and packaging that are related to automobiles.

Acrylate monomers include ethyl acrylate, ethylene-methyl acrylate, methyl methacrylate, 2-chloroethyl vinyl ether, 2-hydroxyethyl acrylate, hydroxyethyl methacrylate, butyl acrylate, trimethylolpropane triacrylate (TMPTA). The preparation of acrylate polymers usually involves the treatment of acrylic acids with the corresponding alcohol in the presence of a catalyst.

Properties of Different Types of Acrylates

Polyacrylic Acid

Polyacrylic acid (or 2-propenoic acid) is the simplest acrylate polymer $(C_3H_4O_2)_n$. Polyacrylic acid polymers are usually produced from acrylic acid that is obtained from propene (a byproduct of ethylene and gasoline production). Acrylic acid polymer is used as thickening agent in acidic conditions because it is highly water dispersible and its dispersion in water thickens with the increase of pH.

Polyacrylic acid and related acrylates are regarded as unusual polymers because they absorb water very freely. They soak up water in multiples of their own weight with no difficulty. This means that even as little grams of the polymer can absorb a cup of liquid, lock it in and hold it up tightly. Such polymers with these characteristics are termed as "super-absorbers". Therefore, it has found usefulness in diapers, especially baby diapers. In diapers, polyacrylic acid absorbs all the liquids that little babies excrete and deposit in it. The super-absorbency of polyacrylics is advantageous in that way that they are less messy and more hygienic because the mess is locked up in the polyacrylic acid polymer and babies feel comfortable while their parents take their time to change the diapers. That is why babies that use diapers of this sort do not get unpleasant skin rashes compared to those using napkins.

Acrylic acids form corresponding esters when reacted in the presence of an alcohol. Acrylic acid and its associated acrylates play important roles in the manufacturing of plastics, diapers, adhesives, floor polishes, nail varnishes/polish, paints, coatings and adhesives of various types.

With its film-forming characteristics, it serves as a film-forming agent that is used to obtain continuous films on the nail, skin or hair. These characteristics enable them to be used in various applications including eyeliner, liquid makeups, mascaras, nail polish, sunscreens, lipsticks and skin care products (creams, lotions, etc).

They also find major applications in coatings and paints such as solvent-born coatings, emulsion paints, interior and exterior water-based paints, and printing inks for applications that need quick drying rates such as auto based lacquers and industrial coatings. They are also used in pressure sensitive adhesive formulations—from low adhesion that are barely tacky to very high tack that can bond permanently to substrates, textiles, automotive products, leather finishing, tape adhesives high-temperature-resistant and oil-resistant elastomers. They can also be used as plasticizer to improve the plasticity of brittle and rigid plastics that are largely used in the automotive industry for dampers, hoses, gaskets and seals that usually function under long-term exposure to hydrocarbon oils and elevated temperatures.

Methacrylate

Methacrylate is often used as a generic term for methacrylate and acrylate. Methacrylate refers to derivatives of methacrylic acid or methyl ester of acrylic acid. These derivatives include the acid form, $CH_2C(CH_3)CO_2H$ or $(C_4H_6O_2)$, the salts such as $CH_2C(CH_3)CO-_2$ Na+ and the esters (e.g. $CH_2C(CH_3)CO_2CH_3$ or methyl methacrylate) and their corresponding polymers.

Methacrylates can form polymers easily because their double bonds are usually very reactive. The polymers exhibit a characteristic water solubility which is pH-dependent. This behavior makes them useful as coating agents in drug delivery systems. It is mostly used as a vehicle for obtaining amorphous systems and solid dispersions, such as nanoparticles, micro particles and microspheres to increase the solubility of low solubility drugs.

Poly(methacrylate) hydrogels have found useful applications in medical devices, especially for ocular applications (e.g. intraocular lenses and contact lenses), and as cell delivery and drug delivery systems. They have found potential applications as intelligent hydrogels for biomimetic sensors, controlled drug release, analyte-sensitive materials, intelligent polymeric membranes and micro-fluidic devices.

Poly(Methyl Methacrylate)

Poly(methyl acrylate) (PMA) is produced with the methyl acrylate (MA) monomer, also called as methyl prop-2-enoate, methyl propenoate or methoxycarbonylethylene. However, poly(methyl methacrylate) (PMMA) is synthesized by polymerization of methyl methacrylate (MMA), which is an acrylate monomer in which one carboxylic acid hydrogen and vinyl hydrogen are replaced by a methyl group.

Methylacrylate is produced by the reaction of esterification with methanol under acid catalysis (acidic ion exchangers, p-toluene sulfonic acid, sulfuric acid) or by pyrolysis of methyl lactate in the presence of ethenone (ketene), the debromination of methyl 2,3-dibromopropanoate with zinc or dehydration of methyl lactate over zeolites as well as vapor-phase oxidation of nickel tetracarbonyl-catalyzed hydrocarboxylation of acetylene and propene or 2-propenal with oxygen in the presence of methanol with carbon monoxide which also yields methyl acrylate. Reacting methyl formats with acetylene in the presence of a transition metal catalyst can also yield methyl acrylate.

PMMA is a linear chain polymer that is characterized by its hydrophobicity, relatively superior light admittance (transparency), amorphous disposition at a molecular level, and glassy feature at room temperature that makes it a strong, hard and clear plastic under the "Plexiglas" trade name. However, it can also become a white rubber at room temperature. The hardness or softness is determined by the numbers of methyl groups present which determines chain mobility and hence the softness and flexibility (i.e., the extent of the free movement of the polymer chains around each other). The softer the polymer, the better the mobility and pliability. Therefore, a polymer will be soft or hard depending on its easy or difficult movement.

MMA is used as the monomer resin in dental materials, in windscreen repair kits and as bone cement for fixing prosthetic devices in orthopedic surgery where it provides a buffering effect, absorbing shock from forces acting upon the bone because of its optimal ability to efficiently distribute stresses and interface strain energy. In its pre-processed form, it is usually a colorless liquid characterized by an acrid odor. It is mainly produced to make acrylate fiber, which is used in weaving synthetic carpets. It also serves as a reagent in the synthesis of various pharmaceutical intermediates. PMMA is used in the preparation of 2-dimethylaminoethyl acrylate by trans-esterification with dimethyl aminoethanol. It is used as a co-monomer with a variety of vinyl and acrylic monomers in the polymerization of various polymers. When using MA as co-monomer, the resulting acrylates are more brittle and harder than those with homologous acrylates. It is also a good dienophile.

PMMA is a non-degradable polyacrylate, which is the most commonly utilized non-metallic implant material in orthopedic surgery and it is also used as an essential ingredient in making dentures. PMMA can be applied as a bone fixing material in the implanting of orthopedic prosthetic materials for shoulders, knees and hips repairs. PMMA-based bone cements can be added to ceramics or bioactive glasses to impart mechanical properties and modify the curing kinetics and as well as loading some dosages of antibiotics in the PMMA cement to decrease the risk of related prosthesis infection. Due to its outstanding bio- and hemo-compatibility and ease of modification, PMMA is usually used in several medical devices, including dialyzers and blood pumps. The optical properties make it a good material for hard contact lenses and implantable ocular lenses. PMMA also offers coloring and physical properties that are beneficial for denture fabrication.

PMMA can be introduced into cosmetic products as beads (porous and non-porous beads) to enhance its performance. The non-porous beads of PMMA impart very high resistance to organic solvents, which allows its usage in nail varnish (mattifying and hardening effects). It has a supple structure and low oil absorption making it a good binder for pressed powders. In addition, they are absolutely transparent on the skin top giving rise to an excellent volume in mascaras, velvet touch effect and ball-bearing effect. Due to their non-porous nature, they do not absorb moisture and sebum neither do they dry the skin. Thus, it makes them ideal candidates for normal, dry and sensitive skin. They are versatile compounds, which can provide soft feel on the skin, optical effects and unique textures to achieve innovative formulations for skin care and make-up.

The porous beads of PMMA are cross-linked and have very high specific area that facilitates the enhancement of absorption of sebum and sweat, occasioning a matte and dry influence on the skin (preventing the sticky skin feel after the application) and optically suppressing makeup smudging.

Polyacrylamide

Polyacrylamide (PAM) readily dissolves in water. Therefore, this polymer is used industrially in many applications where it serves such purposes. In addition, the cross-linked form of polyacrylamides can also absorb water to form gels without dissolving. The absorbed water in the formed gels of cross-linked polymer renders them soft and thus suitable for use in biomedical products such as soft contact lenses.

Another application of PAM gels include soil conditioner to stabilize soil aggregates as well as flocculate suspended particles and furrow irrigation where it reduces erosion and runoff while improving soil and water quality and water-use efficiency. They are also used to reduce surface sealing and crusting as well as erosion in certain agriculture and sprinkler irrigation systems in a cost effective way. PMA is also used to stabilize steep slopes in construction, highway cuts, and other disturbed soils in geo-construction.

Polyacrylonitrile

Polyacrylonitrile (PAN) is a derivative of acrylic acid where the carboxylic acid group is replaced by the related nitrile group and presents the molecular formula $(C_3H_3N)_n$. PAN is a semi-crystalline polymer and though it is a thermoplastic, it does not melt under normal conditions. Its degradation occurs before it melts and its melting is usually observed at above 300°C. When this polymer is heated up to 200°C, it can be transform into a rigid structure with a release of energy, an occurrence referred to as cyclization. Above the cyclization temperature, this polymer undergoes oxidation, a scenario that confers non-flammability on it. If PAN is heated to above 1000°C under inert atmosphere, it gives a product with a percentage of carbon greater than 90%. This property makes it usable in carbon fiber production.

Some major characteristic properties that render PAN special compared to other polymers include the fact that it is the most resistant polymer to degradation by ultraviolet rays and sunlight and it is highly chemically unreactive and resistant to most organic acids and solvents. In addition, this polymer can be attacked only by concentrated solutions of alkalis and highly polar liquids. PAN in the form of acrylic fiber is resistant to breakages, is soft and comfortable, possesses high thermal insulation properties, and it does not melt maintaining its morphological structure when it is heated. This last property of PAN is utilized for the production of insulation fibers, flame retardant fibers, carbon fiber and blankets for filtration of hot gases.

Due to its special properties such as stability to UV degradation, low density, high strength and modulus of elasticity, thermal stability, non-fusibility and chemical resistance, PAN is an important polymer for a wide range of applications. Thus, it is used in textile applications due to its wool-like characteristics, in the making of carbon fiber, filtration membranes, fibers for cement reinforcement, sails for yachts, and awning fabrics in outdoor applications, specialist fibers for acoustic and thermal insulation, anti-flame fibers and in the production of felts for hot air filtration systems.

Copolymers containing polyacrylonitrile, which are flame-retardant, are used as fibers to make knitted clothing, like socks and sweaters, as well as outdoor products like tents. Poly(acrylonitrile-*co*- butadiene-*co*--styrene) (ABS) and Poly(styrene-*co*-acrylonitrile) (SAN) are used as plastics. ABS is a lightweight and very strong polymer. It is so light and strong enough to be used in the production of automobile body parts. Thus, this material renders automobiles' lighter using less fuel and thus polluting less.

The strength of ABS emanates from the nitrile groups that are present in its acrylonitrile units. The nitrile groups are very polar; hence, they can be attracted to one another.

This occurrence permits opposite charges on the nitrile groups to stabilize one another. This strong attraction holds ABS chains very tightly, thus, it makes the material very strong. Furthermore, the rubbery polybutadiene present in it makes ABS a tough polymer.

Ethyl Acrylate

Ethyl acrylate (EA), also named as ethyl propenoate, ethyl prop-2-enoate, acrylic acid ethyl ester or ethyl propenoate, presents the molecular formula $(C_5H_8O_2)_n$. Ethyl acrylate is the ethyl ester of acrylic acid. It is a colorless liquid with a distinctive unpleasantly pungent odor and it is mainly used in the production of paints and textiles, especially non-woven fibers. EA can also be used as a reagent during the synthesis of various pharmaceutical intermediates.

Ethyl acrylate is prepared by acid-catalyzed esterification of acrylic acid that is usually produced by oxidation of propylene. It can also be prepared from ethanol, carbon monoxide and acetylene by a Reppe reaction. EA is used in the manufacture of polymers including rubber, plastics, denture material and resins. It is used as a reactant for alkyl acrylates (acrylic esters) by trans-esterification with higher alcohols through basic or acidic catalysis.

Other acrylate monomers: 2-ethyl hexyl acrylate (obtained from 2-ethylhexanol) is used in pressure-sensitive adhesives; cyclohexyl acrylate (obtained from cyclohexanol) is used in automotive clear lacquers; 2-hydroxyethyl acrylate (obtained from ethylene glycol), which is cross linkable with di-isocyanates to form gels, is used with long-chain acrylates as co-monomer in comb polymers for the reduction of the solidification point of paraffin oils; 2-dimethylaminoethyl acrylate (obtained from dimethylaminoethanol) is used in the preparation of flocculants for sewage clarification and production of paper.

Ethyl acrylate is a reactive monomer. Therefore, it is useful in the enhancement of the cleaning effect of liquid detergents, used as gastric juices tablet covers, in paper, paints and adhesives, textile and leather auxiliaries together with cosmetic and several other pharmaceutical products. EA is also used as a flavoring agent.

2-Ethyl Hexyl Acrylate

2-Ethyl hexyl acrylate, also named as 2-ethyl hexyl prop-2-enoate or abbreviated as 2-EHA with the molecular formula $(C_{11}H_{20}O_2)_n$, is a chemical compound that belongs to the acrylates' family. It is a colorless liquid, which has a sweet odor and can be prepared by the esterification of racemic 2-ethyl hexanol and acrylic acid in the presence of the hydroquinone polymerization inhibitor with a strong acid such as methane-sulfonic acid by reactive distillation using the toluene azeotroping agent.

2-Ethyl hexyl acrylate easily and readily polymerizes. The polymerization of this monomer can be initiated by heat, peroxides, light or even by contaminants. In the presence of strong oxidants, 2-EHA reacts violently with the tendency to form explosive mixtures when exposed to air at temperatures above 82°C (180°F). The chemical, physical and toxicological properties, however, can be greatly modified by additives or stabilizers. 2-Ethylhexyl acrylate is one of the most important base monomers for the production of acrylate adhesives.

Hydroxyethyl Methacrylate

Hydroxyethyl methacrylate (HEMA), also named as 2-hydroxyethyl methacrylate, 2-hydroxyethyl-2-methylprop-2-enoate, glycol methacrylate, glycol monomethacrylate, 2-(methacryloyloxy)ethanol or ethylene glycol methacrylate, presents the molecular formula $(C_6H_{10}O_3)_n$.

HEMA is a colorless and viscous liquid, which can readily polymerize and used as monomer for the production of various types of polymers. It is hydrophobic in nature; nevertheless, in the form of hydrogel, if exposed to water, it swells as a result of hydrophilic pendant group. HEMA is capable of absorbing from 10 to 600% of water relative to its dry weight depending on its chemical and physical structure. For this reason, it is usually used in the production of soft contact lenses.

When treated with polyisocyanates, HEMA makes a cross-linked polymer, an acrylic resin that is a useful component in some paints. HEMA can also be used as a matrix with 40 nm silica particles for 3D glass printing. It is one of the several polymeric hydrogels that are biocompatible with adjustable mechanical properties including resistance to crack propagation. As the first hydrogel ever used for biomedical application, it has been utilized in the production of stents and catheters. Its porous structure eases the diffusion of fluids including oxygen making it suitable for fabrication of hydrophilic contact lens and artificial cornea prostheses.

Poly(2-hydroxyethyl Methacrylate)

Poly(2-hydroxyethyl methacrylate) (PHEMA) is one of the major biomaterial components of most soft contact lenses and is also part of intraocular lenses because of their inherent optical properties. This is due to their hydroxyl groups (–OH groups) being free. PHEMA hydrogels contains very high amounts of water, also posing hydrophilic properties, facilitating the diffusion of oxygen and solutes. That is why oxygen can gain access to the eyeballs through its gel aqueous phase without any hindrance in contact lenses. Lenses that are produced with PHEMA as one of the major constituting materials. Thus, they are widely used because the polymer has a unique blend of good stability to varying conditions of pH and mechanical properties, ionic strength and temperature. Considering its suitability for ocular lenses, PHEMA has exhibited excellent biocompatibility. Hydrogels made from PHEMA and its co-polymers have been critically employed for biomedical uses and considered for controlled drug delivery applications. PHEMA gels have been used in several attempts to reconstruct nasal cartilages, female breasts, as artificial corneas and as wound dressings.

Butyl Acrylate

Butyl acrylate (BA), also called as butyl prop-2-enoate, n-butyl acrylate, butyl ester of acrylic acid or butyl-2-propenoate, presents the molecular formula $(C_7H_{12}O_2)_n$. BA can be synthesized in various different reactions. Thus, 1-butyl alcohol, acetylene, carbon monoxide, hydrochloric acid and nickel carbonyl can react to yield butyl acrylate. Other chemical route in the synthesis of BA comprises the reaction of methyl acrylate with butanol or acrylic acid.

Butyl acrylate is used in coatings, paints, adhesives, sealants, plastics, textiles, caulking materials and fuel. Its derivatives could be used to improve the impact strength of other acrylic plastics, by incorporating them as impact modifiers.

Trimethylolpropane Triacrylate

Trimethylolpropane Triacrylate (TMPTA), also named as 2,2-Bis[(acryloyloxy)methyl] butyl acrylate, presents the molecular formula $(C_{15}H_{20}O_6)_n$. TMPTA is a tri-functional monomer, which is used in the production of adhesives, plastics, inks, anaerobic sealants and acrylic glue. Its usefulness stern from its low volatility and fast cure rate. It possesses good water, chemical, weather and abrasion resistance. The end products of TMPTA include hardwood floors, alkyd coatings, concrete polymers, compact discs, dental polymers, letterpress, lithography, automobile headlamps, elastomers, acrylics, screen printing and different plastic components for medical industry.

Sodium Polyacrylate

Sodium polyacrylate (NaPA), also named as poly (sodium prop-2-enoate), presents the molecular formula $(C_3H_3NaO_2)_n$. NaPA is a sodium salt of polyacrylic acid with broad application in consumer products, and structural engineering. Though it is soluble in water, when processed into a chemically cross-linked polymer hydrogel, it would swell greatly in water. This occurs as a result of the sodium carboxylate salt present in water and the interactions of the hydrogen bonds between water and the cross-linked polymer. This super osmotic process of water absorption tagged this material as a super absorbent (also known as water lock) polymer (SAP) because it is capable of absorbing up to 100–1000 times its weight when exposed to water. When this polymer is blended with clay, it renders improved and controlled swelling properties.

Sodium polyacrylate is an anionic polyelectrolyte-based polymer comprising of carboxylic groups that are negatively charged in the main chain. There are other salts neutralized polyacrylic acids available in the market including potassium, lithium and ammonium. However, sodium neutralized polyacrylic acids are the most common form used in industry.

NaPA and other derivatives of polyacrylic acid have varied commercial and industrial applications: animal care (pet pads, horse urine odor absorbing materials, drown-free water source for feeder insects); topical health (paper and disposable diapers, sanitary wares and napkins); industry (waste liquid control, drilling fluids, concrete protections). environmental protection (geo-textiles for road constructions, anti-flood bags, excreta collection bags and containers); structural engineering application of concrete structures with sodium polyacrylate serving as an internal curing agent and exhibiting enhanced durability and strength performance; biomedical field for drug delivery applications in ocular, nasal, buccal, gastrointestinal, epidermal and transdermal drug delivery system due to their biocompatibility and similarities with living tissues in the pharmaceutical industry; other products such as wire and cables, water blocking, artificial snows, hot/cold gel packs, urine bags, toys, thickening agents, fragrance carriers, fire-retardant gels, anti-fogging packing materials and waterbeds.

Poly(N-Isopropylacrylamide)

Poly(N-isopropylacrylamide) (PNiPAAm) has gained significance as injectable polymer in cell and drug delivery due to the unique physicochemical properties. PNiPAAm undergoes phase separation resulting in the formation of an opaque hydrogel in response to a temperature above 32°C. This thermoresponsive behavior is as a result of strong hydrogen bonds between water molecules, the polymer and the specific molecular orientations of the bonds due to the molecular structure of this polymer.

N-Butyl Acrylate

N-Butyl Acrylate (BA) is a soft monomer with low *Tg* and regarded as the largest-volume acrylate ester used in the preparation of all styrene acrylic copolymers, acrylic and vinyl acrylic. The other major acrylate esters include 2- ethylhexyl acrylate (2-EHA), ethyl acrylate (EA), and methyl acrylate (MA). Butyl acrylate is used as a soft-monomer to improve toughness and low temperature. The major applications for BA are in coatings and paint within the automotive and architectural field, sealants, adhesives and intermediates. Other areas of BA application include leather finishes, inks, caulks, textiles and papers.

Another important and growing usefulness of BA is in thermoplastic ethylene acrylate co-polymers (EAC). They are used as processing aid and impact modifier in thermoplastics to improve properties such as flexibility, toughness, molding characteristics, part appearances and low temperature properties. End use applications include adhesives, packaging and multilayer films. BA showcases superior photostability and is a preferred monomer where weatherability, with emphasis on low temperature applications especially at high altitudes, and sunlight resistance are required.

The combination of butyl acrylate with other polymerizable acrylate ester monomers allows the fabrication of copolymer compositions. This way, the performance of co-polymers can be tailored to meet a wide range of end-use requirements as it has the capacity to balance softness and hardness, block resistance and tackiness, low temperature flexibility, durability and strength, and other important properties required of products to meet end-use goals.

Copolymers

A copolymer is a polymer that is made up of two or more monomer species. Many commercially important polymers are copolymers. Examples include polyethylene-vinyl acetate (PEVA), nitrile rubber, and acrylonitrile butadiene styrene (ABS). The process in which a copolymer is formed from multiple species of monomers is known as copolymerization. It is often used to improve or modify certain properties of plastics.

A homopolymer is a polymer that is made up of only one type of monomer unit. The difference in the constitution of a copolymer and a homopolymer is described through figure.

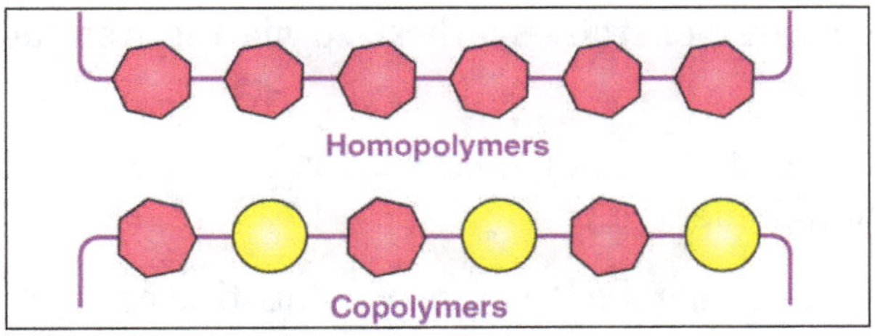

Copolymers are categorized based on their structures. Those containing a single chain are known as linear copolymers whereas those containing polymeric side chains are called branched copolymers.

Classification of Copolymers

Types of Linear Copolymers

Linear copolymers can be further classified into several categories such as alternating and statistical copolymers. This classification is done based on the arrangement of the monomers on the main chain.

Block Copolymers

- When more than one homopolymer units are linked together via covalent bonds, the resulting single-chain macromolecule is called a block copolymer.
- The intermediate unit at which the two homopolymer chains are linked is called a junction block.

- A diblock copolymer contains two homopolymer blocks whereas a triblock copolymer contains three distinct blocks of homopolymers.
- An example of such a polymer is acrylonitrile butadiene styrene, commonly referred to as SBS rubber.
- An illustration describing the structure of a block copolymer which is made up of the monomers 'A' and 'B' is provided below:

—B—B—B—B—A—A—A—A—

Statistical Copolymer

- Statistical copolymers are the polymers in which two or more monomers are arranged in a sequence that follows some statistical rule.
- Should the mole fraction of a monomer be equal to the probability of finding a residue of that monomer at any point in the chain, the entire polymer is then known as a random polymer?
- These polymers are generally synthesized via the free radical polymerization method.
- An example of a statistical polymer is the rubber made from the copolymers of styrene and butadiene.
- An illustration describing the structure of a statistical copolymer is provided below:

—A—B—A—B—

Alternating Copolymers

- Alternating copolymers contain a single main chain with alternating monomers.
- The formula of an alternating copolymer made up of monomers A and B can be generalized to $(\text{-A-B-})_n$.
- Nylon 6,6 is an example of an alternating copolymer, consisting of alternating units of hexamethylene diamine and adipic acid.
- An illustration describing the general structure of these polymers is provided below:

—A—B—A—B—

Periodic Copolymers

These polymers feature a repeating sequence in which the monomers are arranged in a single chain. An illustration of the structure of a periodic copolymer made up of monomers A and B is provided below:

$$\left[A-B-A-B-B-A-A-A \right]_n$$

Gradient and Stereoblock Copolymers

The single-chain copolymers in which the composition of monomers gradually changes along the main chain are called gradient copolymers. If the tacticity of the monomers varies with different blocks or units in the polymer, the macromolecule is known as a stereoblock copolymer.

Branched Copolymer

As the name suggests, a branched copolymer is a polymer in which the monomers form a branched structure. Some important types of branched copolymers include star, comb, grafted, and brush copolymers. A star copolymer contains several polymeric chains that are attached to the same central core.

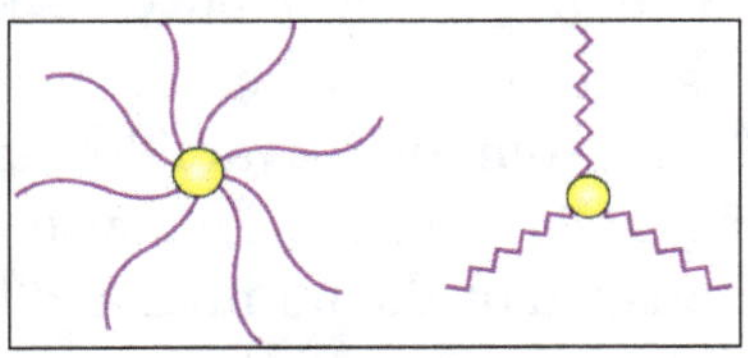

The structures of different types of star-shaped copolymers are illustrated above. They consist of a multifunctional center to which three or more polymer chains are attached.

Graft Copolymers

Branched copolymers featuring differently structured main chains and side chains are known as graft copolymers. An illustration detailing the structure of a graft copolymer made up of monomers A and B is provided below:

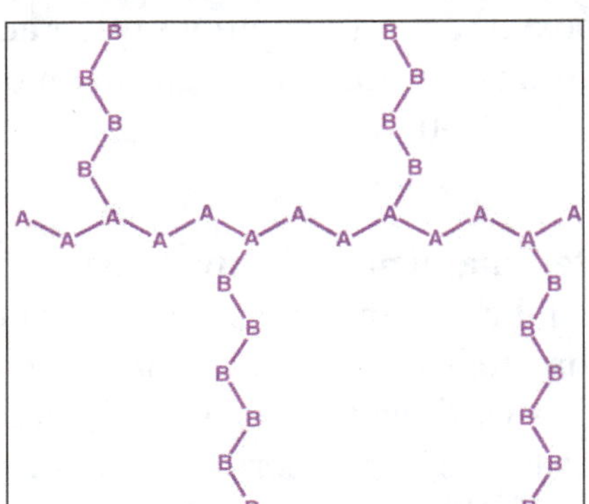

The main chain or the side chains of these polymers can be copolymers or homopolymers. High impact polystyrene is an important example of a graft copolymer. They can be synthesized from free radical polymerization.

Polyamides

Polyamide is any polymer (substance composed of long, multiple-unit molecules) in which the repeating units in the molecular chain are linked together by amide groups. Amide groups have the general chemical formula CO-NH. They may be produced by the interaction of an amine (NH_2) group and a carboxyl (CO_2H) group, or they may be formed by the polymerization of amino acids or amino-acid derivatives (whose molecules contain both amino and carboxyl groups).

Broadly defined, the polyamides include proteins and peptides which are naturally produced polymers consisting of amino-acid repeating units. (In molecular biology the amide linkage is usually referred to as the peptide bond.) More narrowly defined, the polyamides are an important group of industrially produced synthetic polymers. The most important is nylon, actually an extremely versatile class of polymers that are made into indispensable fibres and plastics. Another class of polyamides made into fibres is the so-called aramids, or aromatic polyamides—amide polymers that contain phenyl rings in their repeating units. Some nylons also contain phenyl rings, so the two classes are actually distinguished by the number and frequency of the rings. When 85 percent or more of the amide groups are linked directly to phenyl rings, the polyamide is classified as an aramid. When the rate is lower than 85 percent, the polymer is classified as nylon.

Polyesters

Polyesters are one of the most important and most used polycondensation polymers and are derived from dicarboxylic acids (sometimes other acid types) and diols. Polyester is a polymer class containing ester functional group on polymeric main chain. Polyester term is usually used for polyethylene terephthalate (PET), despite the numerous polyester forms are present.

Recent studies have indicated that using of polyester polymer processed into various forms, e.g., fibers, filament, fabric, composites, resins, dendrimers, films, sheets, and membranes in different fields, such as textile, automotive, medical, electronic, and construction applications, attracts worldwide interest. Polyester is also extensively used as packaging materials, such as bottle/containers. The polymerization of polyester could be carried out as polycondensation, ring-opening polymerization, and polyaddition processes. Furthermore, polyester could be recycled by physical (mechanical) or chemical (hydrolysis, methanolysis, and glycolysis reactions) methods.

Recycled polyester could be used for packaging, construction parts, pipes, tanks, geotextiles, nonwoven, carpets, etc. It is expected to run out of crude oil reserves at World in 2043. Thus, recycling of petroleum-based polymers is crucial. In addition, effectively recycling of polyester will give rise to lessen carbon dioxide emission and thus global warming.

Polyester can be classified into two groups: thermoplastic polyesters and thermoset (unsaturated polyester, polyester resin) polyesters. Thermoplastic polymers could be categorized as linear aromatic polyesters (fiber- and film-forming polyesters), elastomers (block copolyesters), liquid crystal polyester, engineering plastics, aliphatic polyesters, and poly(hydroxyl alkanoates).

Thermoset Polyesters (Unsaturated Polyester)

Unsaturated polyester resins (UPRs) are used in civil/structural engineering applications, ships materials, composites, construction, piping, storage tanks, protective coatings, and automotive paints, which required high strength, ductility properties, and fire resistance. The composite materials composed of unsaturated polyester are produced by some methods, such as lay-up method, pultrusion, filament winding, vacuum bagging and autoclave curing, and liquid molding. Unsaturated polyester resins have advantages due to their chemical resistance, electrical properties, rapid curing, and relatively low prices for using the application areas. However, UPRs are extremely flammable and produce toxic smoke during combustion. Some filler and additives such as alumina trihydrate (ATH), magnesium hydroxide, nickel hydroxide, molybdenum disulphide, nanoclay, antimony oxide, gypsum particles, graphene, and carbon nanotube in the resins and some modifications of the resins with halogens such as bromine or phenolic resin could be used to improve fire resistance of the composites. Furthermore, their stiffness and strength properties were limited in comparison with other thermosetting resins. Fiber-reinforced unsaturated polyesters (UPs) and filler containing UPs were widely investigated to surpass these deficiencies. Applications of unsaturated polyester resins were illustrated in Figure below.

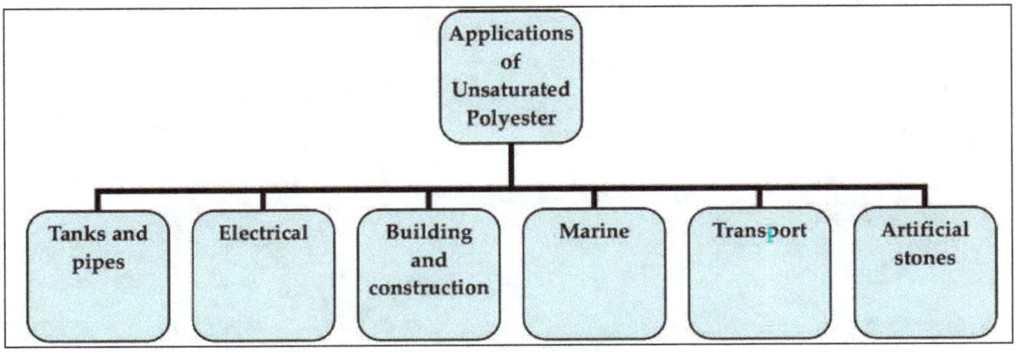

Figure: Applications of unsaturated polyester resins.

Thermoplastic Polyesters

Polyester is the most produced thermoplastic polymer and has many application fields such as especially textile fibers and bottles. Polyester used in textiles is generally polyethylene terephthalate (PET), whose well-known properties are chemical inertness, lightness, good processability, high-melting point, high tenacity, and low cost. Furthermore, recycling of polyester bottles for sustainable textiles recently draws attention in the world and provides diverting waste as bottles from landfills, reducing environmental pollution and reducing carbon footprint and save energy compared to produce virgin polyester. The bottle could be recycled by mechanical and chemical processes. Chemical process is based on the depolymerization of polyester to oligomers or monomers by chemical reaction, while mechanical process is carried out with melting and re-extruding to make fiber.

References

- Organic-polymer, materials-science: sciencedirect.com, Retrieved 09, May 2020
- Properties-and-applications-of-acrylates, acrylate-polymers-for-advanced-applications: intechopen.com, Retrieved 14, June 2020
- Copolymer, chemistry: byjus.com, Retrieved 26, February 2020
- Polyamide, science: britannica.com, Retrieved 18, April 2020
- Polyester-production-characterization-and-innovative-applications, introductory-chapter-introduction-to-polyester-production-characterization-and-innovative-application: intechopen.com, Retrieved 02, January 2020

4

Inorganic Polymers

Inorganic polymers are the polymers that do not contain carbon atoms in their backbone. They have ionic and hydrocarbon bonds in their backbone along with some heteroatom such as oxygen, nitrogen and sulphur. This chapter closely examines the key concepts of inorganic polymers to provide an extensive understanding of the subject.

Inorganic polymers are polymers with a skeletal structure that does not include carbon atoms in their backbone. Inorganic polymers have ionic and hydrocarbon bonds in their backbone along with some heteroatom such as oxygen, nitrogen and sulphur. Inorganic polymers in aqueous solutions are good conductors of electricity, this is because they have high ability to ionize and this makes them better conductors. Inorganic polymers are highly soluble in common organic solvents and water. This is because they have ionic bonds between molecules. Ionic bonds easily dissociate into positive and negative ions in water. Inorganic polymers are widely used in petrochemical industries while others such as silicone rubber are used in construction for window and door seals. Also in electrical engineering inorganic polymers such as silicon rubber is used in wire and cable jacketing and in electrical safety stinger covers Other inorganic polymers such as polydimethylsiloxane are widely used as a versatile ingredient in many skin care and beauty products because of its ability to serve as an anti-foaming agent, skin protectant and conditioner.

Inorganic polymers (IP) are commonly defined as materials constituted by aluminosilicates conventionally designed by activating a solid alkali silicate. Geopolymers are inorganic polymer often regarded as a subset of IP whereby the tetrahedral coordination of the aluminosilicate framework exclusively comprises Si atoms and Al atoms. The geopolymers are environmental friendly, durable, corrosion resistant, and inflammable. The emergence of mesoporous inorganic polymers such as titania and silica-based polymers provides cutting-edge water remediation technologies owing to their photocatalytic, high thermal stability, versatility, reusability, and attrition resistance property; electronic conductivity; high surface area; surface reactivity; structural stability; surface charge; and molecular sieve porosity that allows mass contact.

Although organic polymers are generally inexpensive, it is the physicochemical properties of engineered inorganic polymers that make them interesting for researchers. To date, most inorganic polymer-based engineering material comprises silicon, phosphorus, and nitrogen in their skeletal structures. Much as most polymers made from these atoms can selectively remediate aquatic systems, the crafting of other strategic inorganic polymer composite remains ongoing.

Currently, geopolymer and smart polymer composites are the commonly used inorganic polymers for environmental applications. Various inorganic polymer-based engineering materials have been used to effectively facilitate water remediation by coagulation (e.g., polyferric silicate sulfate), adsorption (polyamine-functionalized MCM-41 mesoporous silica (MS-PEI), and photodegradation.

Inorganic polymer-based engineering material is characterized by easy synthesis at ambient temperatures, multiligand binding, electric conducting, antifouling, and hydraulic properties. Other advantages of inorganic polymer-based engineering material used for industrial wastewater remediation technologies include low water usage, low emission of CO_2, and reduced operational cost.

Geopolymers

Geopolymers are inorganic polymeric materials obtained by mixing of a dry solid (aluminosilicate) with an alkaline solution and other constituents, if necessary. The major constituent is the source material, this one must be rich in silicone (Si) and aluminum (Al), they could be natural minerals such as clays, kaolinite, etc. or red mud, fly ashes, slags, etc. known as "waste" materials. The liquid is usually based on sodium (hydroxide or silicate) or potassium (hydroxide or silicate) with are soluble alkali metals. The time of hardening is very short; in first 4 hours of setting they obtain 70% of the final compressive strength.

The structures of geopolymers consist in three dimensional silico-aluminate links (sialate) of SiO_4 and AlO_4 tetrahedra with share all the oxygen atoms, designed as in Table below. Geopolymer matrix presents an empirical Formula 1, as following:

$$M^{+}_{v}\{-(SiO_2)x-AlO_2-\}_{v} * aH_2O$$

Where: M^+ - Alkali cation of K^+, Na^+ ; v - degree of polymerization; x – Si/Al ratio, a – the amount of water, if necessary (when the alkali activator is solid, a specific amount of water is necessary to create the reaction environment, if the activator is liquid, there is no need of supplementary water). The properties of the geopolymer are highly depending of the value of z (z = 1-15, up to 300), for high value of z (z>3) geopolymers exhibit adhesive and rubbery properties, their structure being a linear 2- D network, when the value of z is low (z ≤ 3) the material presents stiff and brittle properties and a cross-linked 3-D network.

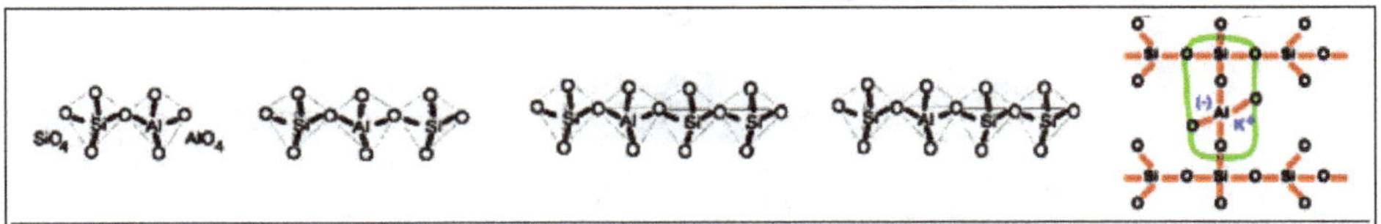

Poly(sialate) -Si-O-Al-O-	Poly(sialate-siloxo) -Si-O-Al-O-SI-O-	Poly(sialate-disiloxo) -Si-O-Al-O-Si-O-Si-O-	Poly(sialate-disiloxo) -Si-O-Al-O-Si-O-Si-O-	Sialate link
Si:Al = 1	Si:Al = 2	Si:Al = 3	Si:Al = 3	Si:Al > 3

Figure: Types of poly(sialates).

The geopolymer result after the exothermic process carried out through oligomers, process known as geopolymerization, is a polymer with very long reticular network, where the specific tetragonal network of aluminates (AlO_4) and groups of silicates (SiO_4) are, also, present. The bond between these tetrahedrons is equilibrated by alkali ions of: K^+, Na^+ or Li^+.

In general, geopolymerization can be divided into three stages, see Figure above. First of them is dissolution, when the solid alumino-silicates material is dissolved because of the water and alkali activator presence. After eliminating a small amount of water, the reorientation starts, now the group atoms take their place in the structure. During the solidification (at 20 Celsius degree or higher approximate 1000 °C) the water is almost totally eliminated and the material shows his final form.

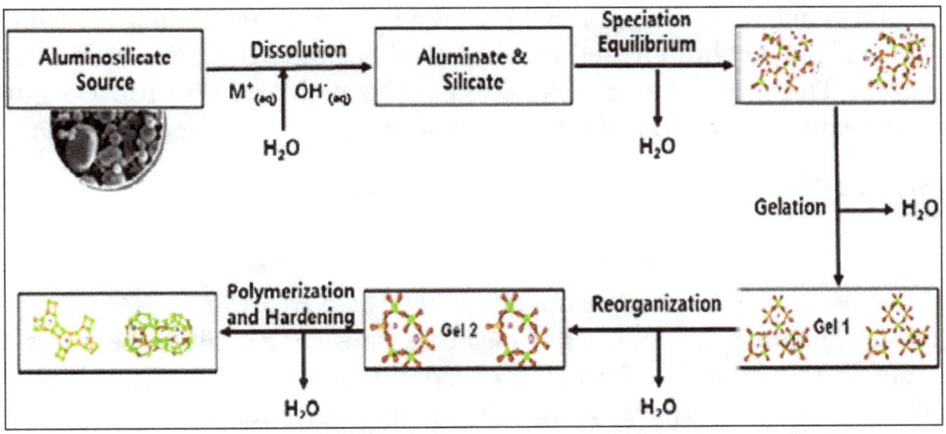

Figure: Stages of geopolymerisation process.

Materials for Geopolymers

Because of the fact that any material rich in aluminum and silicone see Figure below, can be used as raw materials, there are many materials that can be used to create geopolymers. Kaolinite was the first material widely used in geopolymer synthesis after successfully uses of this new material, the scientists start developing new raw materials, such as calcinated clays or industrial waste (e.g. slag, ashes, waste glass, aluminum mine tailings, etc.) and natural silicoaluminates (e.g. pure Al_2O_3–$2SiO_2$ powder, zeolites, etc).

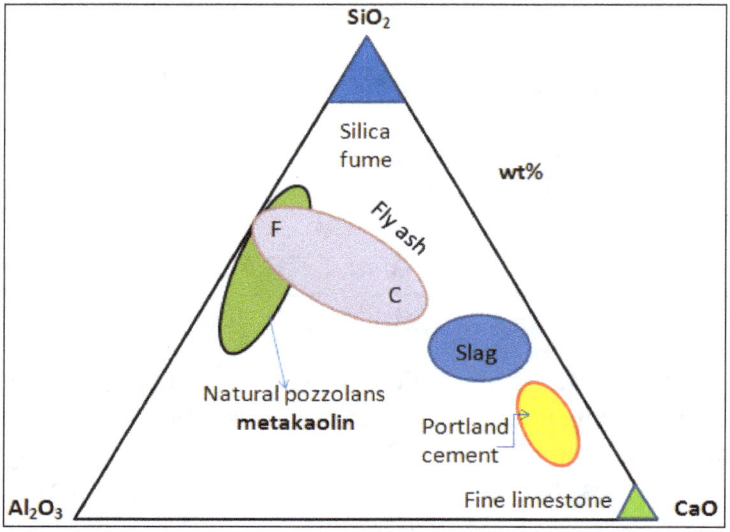

Figure: $CaO–Al_2O_3–SiO_2$ ternary diagram of cementitious materials.

Geopolymers Obtained using Metakaolin

The first use of geopolymers was as building products, this products include fire-resistant chipboard panels, made from a wooden core faced with two coatings and ceramic tiles water-resistant at temperatures lower than 450°C, without firing. At 150°C kaolinite that is one of the clay components reacted with caustic soda. The industrial applications of this kaolinite reaction with alkali began in the ceramic industry with Olsen in 1934. The alumino-silicate kaolinite reacts with NaOH at 100°C - 150°C and polycondenses into hydrated sodalite (a tectoaluminosilicate), or hydrosodalite,

$$Si_2O_5,A_{l2}(OH)_4 + NaOH \Rightarrow Na(-Si\text{-}O\text{-}Al\text{-}O)n$$

Kaolinite Hydrosodalite

Kaolin, after being calcinated at temperatures between 650°C and 750°C, is the main source of metakaolinite. Kaolin is a phyllosilicate, consisting of alternate layers of silica and alumina in tetrahedral and octahedral coordination, respectively. This electrically neutral crystalline layer structure, which is a common characteristic of clay minerals, leads to a fine particle size and plate like morphology and allows the particles to move readily over one another, giving rise to physical properties such as softness, soapy feel and easy cleavage. Kaolinite is the mineralogical term for hydrated aluminium disilicate, $Al_2SiO_5(OH)_4$, the primary constituent of kaolin (40–70%). Other minerals comprising kaolin include quartz, muscovite-like micas and rutile. When kaolin heated to temperature of 650–900 C, it loses 14% of its mass in bound hydroxylions. This heat treatment, or calcination, breaks down the structure of kaolin such that the alumina and silica layers become puckered and lose their long-range order. Resulting from this dehydroxylation and disorder is metakaolinite, a highly reactive transition phase, amorphous material with pozzolanic and latent hydraulic reactivity, suitable for use in cementing applications.

Metakaolin-based geopolymers can be manufactured consistently with a high level of predictability in hardened characteristics, see Figure below, displays compressive strengths and porosities for both metakaolin geopolymers and metakaolin/fly ash mixtures.

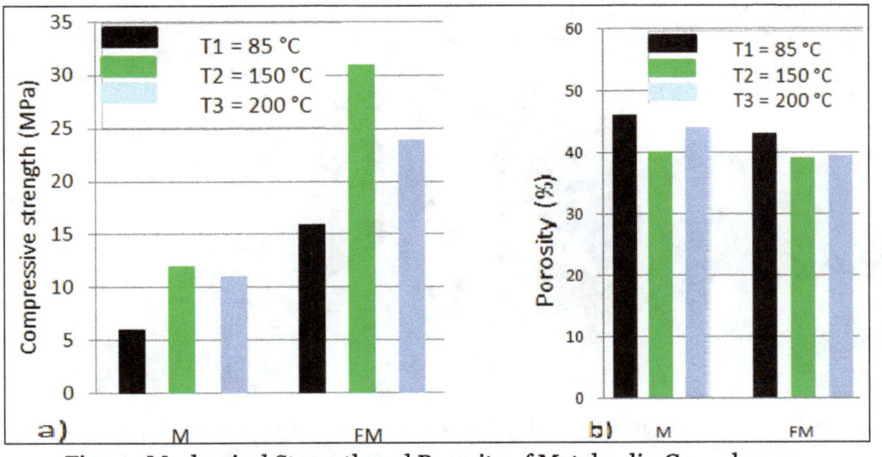

Figure: Mechanical Strength and Porosity of Metakaolin Geopolymers.

Geopolymers Obtained using Ashes

The need of new construction materials with high properties is in a continuous growth and so is the need of reducing of using the natural minerals for this purpose. In this condition the use of ashes in cements or concretes is a great "green" idea. Beside the high chemical, physical and mechanical properties, the positive effect on environment and low emission of CO_2, the economical aspect is also important, playing an important role when this material is chosen. Fly ashes are spherical glassy particles obtained by solidification of coal by-product volatile material. Due to the fusion-in-suspension these fly ash particles are mostly minute solid spheres and hollow ecospheres with some particles even being plerospheres, which are spheres containing smaller spheres.

Fly ash can be divided into two classes based on its source of origin and composition. Class F is a fly ash is obtained from burning of bituminous or anthracite coal and meets this chemical composition $SiO_2 + Al_2O_3 + Fe_2O_3 \geq 70\%$. Class C is normally produced from burning of sub-bituminous or lignite coal and has the following chemical composition $SiO_2 + Al_2O_3 + Fe_2O_3 \geq 50\%$.

Geopolymers Obtained using Red Mud

There are multiple uses of this highly alkaline (pH of the order of 11 or higher) waste generated by aluminum production plants (for 1 tons of aluminum, 2 tons of red mud are deposited) however the reuse rate is very small and this material become a huge problem for the environment. Red mud is a mixing of minerals with complex chemical composition, where iron has the highest concentration followed by aluminum, silica, calcium, sodium and low concentration of rare earth chemical elements, like: Dy,Er, Yd, Nd, Sm, Eu, Gd, Sc, La, Ce, Y, Pr. Bayer Process consist in dissolving of bauxite with (NaOH) at 175° C resulting aluminium hydroxide $Al(OH)_3$ and 50% -70% red mud. Red mud is usually used to create geopolymer cements with low CO_2 emission, small price and excellent quality.

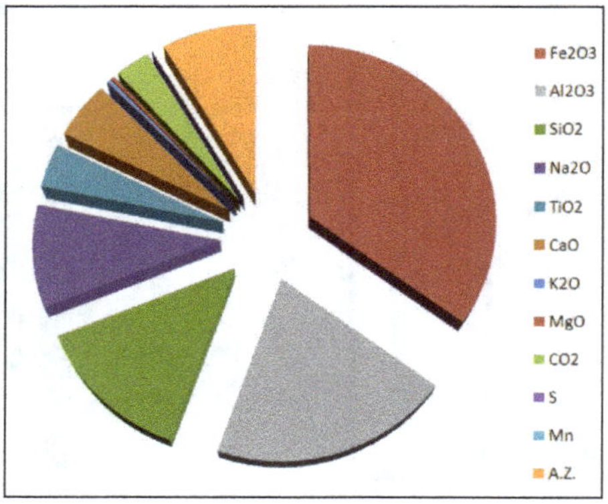

Figure: Typical red mud chemical composition.

Geopolymers Obtained using Slags

Blast furnace slag is a foundry residue result from production of cast iron. In this process, the iron is melted and mixed with different additions in order to eliminate the impurities from metal bath, when the slag is expelled on the bath surface it is fast removed and cooled down in water. The result material, slag, has amorphous structure and chemical composition reached in calcium aluminates and silicates. The slags can be used in manufacture of different type of materials and by geopolymerization is used to create binders, mortar and different types of concretes with good mechanical properties.

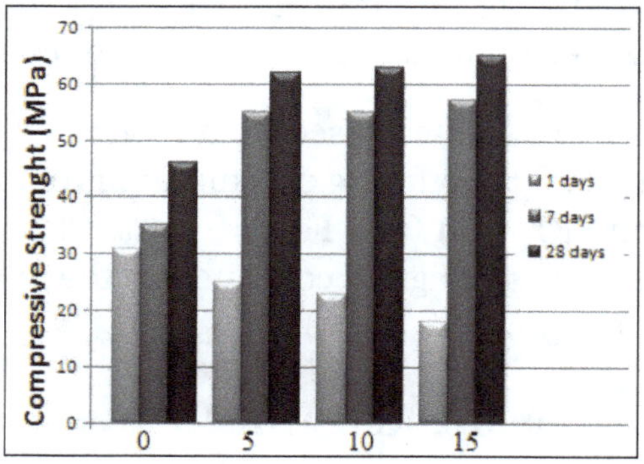

Figure: Compressive strength value for different types of slags (GBFS)-metakaolin (MK) based geopolymers.

Alkaline Activators

Activation of the selected pozzolanic material is the most significant factor in producing a mechanically-sound cementitious material via the geopolymerization process. The activators prompt the precipitation and crystallization of the siliceous and aluminous species present in the solution. OHacts as a catalyst for reactivity, and the metal cation serves to form a structural element and balance the negative framework carried by the tetrahedral aluminum. The initial mechanism of reaction is driven by the ability of the alkaline solution to dissolve the pozzolanic material and release reactive silicon and aluminum into solution. There are multiple activating agents that can be used for geopolymerisation (M-alkali ion): Alkalis, MOH; Strong Salt Acids, M_2SO_4, Aluminosilicates, M_2O * nAl_2SO_3 * (2-6)SiO_2; Aluminates, M_2O * $nAlO_3$; Silicates, M_2O * $nSiO_3$; Weak acid salts, M_3PO_4, M_2SO_3, M_2CO_3, MF.

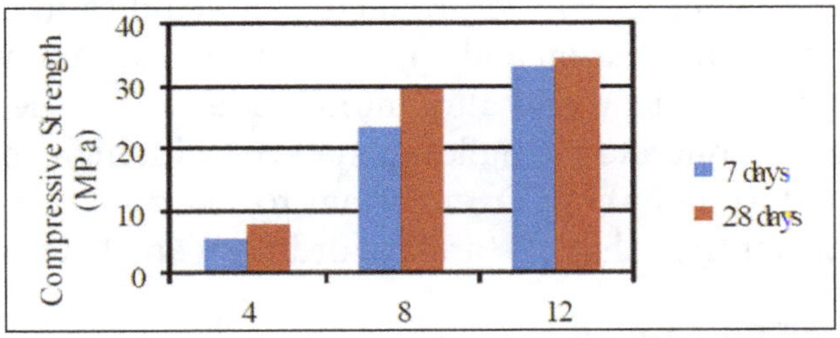

Figure: Compressive strength values depending of NaOH concentration.

Sodium Hydroxide (NaOH) is a geopolymer activator commonly used as alkaline solution. The sodium cations promote better geopolymerisation, compared with potassium cations, due to the fact that they can migrated easy thru the network in the gelatin stage being smaller. The strength value of geopolymers increase with the increasing of NaOH concentration, see Figure above at early stages. Yet, the aged material will be compromised due to excessive OH that modifies morphology in a non-uniform one. These types of activated geopolymers are more stable in sulfates and acids environments by having greater crystallinety. Potassium hydroxide (KOH) activator improves the compressive strength of the geopolymers cements and their porosity. K^+ will offer a denser final structure compared with other activators because of its high reactivity with the alumino-silicate material.

Sodium silicates (Na_2SO_3) are rare single used activator for geopolymers being usually mixed with NaOH to achieve enough activation potential. It is used mostly for its increasing strength and alkalinity properties. The most used geopolymerisation activator is a combination between Na_2SO_3 and KOH with different value ratio.

The Processing Parameters Effect

Because of the fact that, the final properties of the geopolymer material are highly depending of the processing parameters, the effect of different processing factors was studied by many scientists studied the influencing of four of the main factors: curing temperature and molar ratios of SiO_2/Al_2O_3, Na_2O/SiO2 and H_2O/Na_2O as can be observed in the Figure above. They conclude that, by increasing this all of these major factors, the mean value of compressive strength decreases. Also, they observed that, the curing temperature have no significant effect on this properties for short period of time. After passing of a limit value of SiO_2/ Al_2O_3 (2.9 %) and Na_2O/SiO_2 (0.2 %) molar ratio, the system will be saturated with Na+ ions and the formation of oligomers is prevented. The molar ratio of H_2O/ Na_2O presents the highest negative effect by prevention from dissolving of alumino silicate and by high porosity introduced by water withdrawal, presenting high decrees of properties between concentration of 13.75% and 21.75 %.

The Properties of Geopolymers

- Environmental friendliness: GPs, an environmental protective silicon aluminate inorganic material, whose products can significantly reduce the carbon footprint in their preparation and application process. And in the nature world, various containing silicon and aluminum oxide can be used to synthesize GP. For instance, GP concrete is a building material with high application potential. It can reduce the amount of CO_2 emissions to a large extent, and the maximum can be 86% lower than that of standard ordinary Portland cement.
- Rapid hardening and high strength: GP has the characteristics of high early strength and fast setting time. In recent years, researchers have done a lot of research on GP owing to its high mechanical strength. Du revealed that using SiC can obviously enhance the compressive strength of GP composites. When the content of SiC is 10 wt %, the compressive strength of GP composites can reach 155 MPa, which is equivalent to increasing 100% on the basis of unfilled GP.
- High temperature resistance and heat insulation: The oxidation reaction and decomposition reaction of the GP products between 1000-1200 °C are not possible, and they are stable under high temperature. The linear shrinkage at 400 °C and 800 °C are 0.2% to 1% and 0.2% to 2%, respectively, and the original strength above 60% can be maintained. This shows better mechanical strength at elevated temperature.
- Good corrosion resistance and durability: GPs show good stability in acidic solutions and in various organic solvents. In 5 wt.% hydrochloric acid and sulfuric acid solution, the decomposition rate of GP were only 1/13 and 1/12 of it of Portland cement, respectively. GP can form compact structure with high strength and excellent impermeability.
- Strong interfacial bonding ability: At the interface between the traditional Portland cement and the aggregate, the transition zone of calcium hydroxide enrichment and preferred orientation is easy to occur, resulting in weak interfacial bonding force. And for GP, the transition zone does not appear due to the reason that the hydration reaction of calcium silicate does not exist, and the final product is a three-dimensional network of gel, which combined with the aggregate interface.

- Immobilizing toxic metal ions: Mallowc and others believe that metal ions are also involved in the formation of GP structures, so that metal ions in the system can be more effectively fixed.

The Application of GPs

The excellent properties of GPs determine their broad application and development prospects, and the main applications are as follows:

Innovative GP Binder and Sustainable Repair Material

Utilization of GP binder has been extensively investigated for its low shrinkage, high early strength, sulfate attack and corrosion, high resistance to thawing and freezing. Used in civil engineering, it can greatly shorten the stripping time, speed up the template operation cycle, and improve the construction speed. One-part GP mixtures is a new type of geopolymeric binders, which can be used to simplify the process of activating GPs in silicate solution. At the temperature of 950°C, Peng has performed thermal treatment of low-quality kaolin, and the strength of the corresponding binders is up to 47MP, which could reach comparable strength with two-part mix systems. Nasvi revealed that GP cement is suitable for storage wells under deep down-hole stress conditions and carbon capture owing to its good acid resistance, high mechanical strength, durability and low permeability. GP cement is regarded as potential replacement of Portland cement. This is mainly based on the properties and sustainability criteria.

GP-Type Coatings

The GP can be connected into the network structure of film material through the aluminum tetrahedral and silicon tetrahedral, so as a kind of inorganic coating, with non-toxic environmental protection, low cost, fire proof and water proof etc. Various proportions of β-SiC were used to prepare the GP-type paints by A. Khosravanihaghighi, and the results show that, by increasing the proportion of nano-SiC, the porosity is reduced, and the corrosion resistance and wear resistance are improved. In addition, reflective heat insulation GP coating was synthesized from metakaolin and sodium silicate solutions as main initial raw material, and talcum powder, hollow glass microspheres and sericite powder are added as fillers. The properties of two hybrid type GP concrete composed of alkaline-activated metakaolin and fly ash and it was treated as protective coatings against chloride-induced corrosion in its application.

Novel Ceramic Materials

Compared with the hydrothermal and the sol-gel methods, GP technology provides a new method for the direct synthesis of final structural ceramics. GPs can form a structure comparable to those of high temperature sintered ceramics, and the low temperature GP ceramic is convenient for forming products of various complex shapes and has good overall performance.

Sealing Materials for Industrial Toxic Waste Residue

Harmful heavy metals can be produced from modern industry, agriculture, disposal of waste and others. GP technology can transform industrial solid waste containing aluminosilicates into useful products thanks to the flexibility and capability to immobilize and stabilize the wastes inside the GP network. Metals such as Co, Cu, Pb, Cd, Ni, Zn, Pd, As, Ra and U can be incorporated in the 3 dimensional GP network decreasing the mobility of the heavy ions through metal hydroxide precipitation, ion substitution or physical encapsulation.

GP Based Self-Cleaning Concrete Materials

Severe air pollution is mainly caused by industrial production, natural disasters, engineering construction and automobile exhaust, etc. Thus building facilities will become dirty when exposed to the pollute air. This problem has forced the industries to develop alternative materials for supporting the sustainable development. GP based self-cleaning concrete material is green alternative for solving the problem owing to their unique properties. And the self-cleaning performance of GP based concrete is studied by introduction of photocatalytic materials such as ZnO and TiO_2.

GP Biological Materials

GP composite has the advantages of non-toxic, biocompatibility, good drug carrying capacity and sufficient mechanical strength, etc., which can be used as oral drug delivery carrier. Metakaolin-based Na GP has been reported as an appropriate carrier for the powerful synthetic the sedative drug zolpidem and opioid analgesic drug fentanyl. Furthermore, the ion exchange property of GP has been developed into germicidal material with potential application prospect of water purification, and it has a great application prospect in water-scarce countries.

The Applications of other Fields

GPs have a wide variety of applications in other fields owing to its excellent properties. For example, there are some of the following:

- Thermal shock refractories.
- Fire-resistant materials, thermal insulation, foams.
- Environmentally friendly adhesive for wood based panels and modern biotechnology.
- High-tech carbon fiber materials used inside and outside of aircraft.

- Dielectric materials for electronic packaging and molecular sieve catalysts material.

Silicones

Silicone, also called polysiloxane is any of a diverse class of fluids, resins, or elastomers based on polymerized siloxanes, substances whose molecules consist of chains made of alternating silicon and oxygen atoms. Their chemical inertness, resistance to water and oxidation, and stability at both high and low temperatures have led to a wide range of commercial applications, from lubricating greases to electrical-wire insulation and biomedical implants (such as breast implants).

Figure: Caulk-Silicone caulk being dispensed from a caulking gun.

Composition, Structure, and Properties

The silicones differ from most industrial polymers in that the chains of linked atoms that make up the backbones of their molecules do not contain carbon, the characteristic element of organic compounds. This lack of carbon in the polymer backbones makes polysiloxanes into unusual "inorganic" polymers—though in most members of the class two organic groups, usually vinyl (CH_2), methyl (CH_3), or phenyl (C_{6H5}), are attached to each silicon atom. A general formula for silicones is $(R_2SiO)_x$, where R can be any one of a variety of organic groups.

The most common silicone compound, poly-dimethylsiloxane, can illustrate the central characteristics of the class. The starting material is metallic silicon, which is obtained from silica sand. Silicon is reacted with methyl chloride (CH_3Cl) over a copper catalyst, forming dimethyldichlorosilane ($[CH_3]_2Si[Cl]_2$). By reacting this compound with water, the chlorine atoms are replaced by hydroxyl (OH) groups. The resultant unstable compound, silanol ($[CH_3]_2Si[OH]_2$), polymerizes in a condensation reaction, the single-unit molecules linking together to form poly-dimethylsiloxane with concomitant loss of water. The dimethylsiloxane repeating unit of the polymer has the following structure:

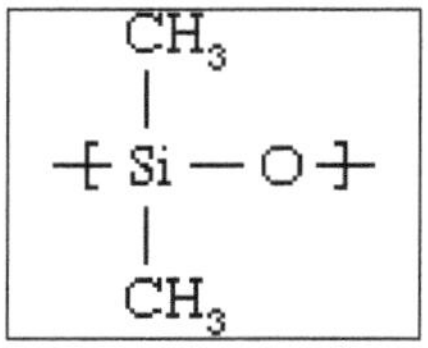

Siloxane molecules rotate freely around the Si-O bond, so, even with vinyl, methyl, or phenyl groups attached to the silicon atoms, the molecule is highly flexible. In addition, the Si-O bond is highly heat-resistant and is not readily attacked by oxygen or ozone. As a result, silicones are remarkably stable, and they have the lowest glass-transition temperature (the temperature below which the molecules are locked in a rigid, glassy state) and the highest permeability to gases of any polymer. On the other hand, the Si-O bond is susceptible to hydrolysis and attack by acids and bases, so silicone plastics and rubbers are relatively weak and readily swollen by hydrocarbon oils.

Applications

Polysiloxanes are manufactured as fluids, resins, or elastomers, depending on the molecular weight of the polymers and the degree to which the polymer chains are interlinked. Nonvulcanized, low-molecular-weight polysiloxane fluids are exceptionally stable to decomposition by heat, water, or oxidizing agents and are good electrical insulators. They make excellent lubricants and hydraulic fluids, as well as emulsions for imparting water repellency to textiles, paper, and other materials. Silicone resins are used in protective coatings and electrically insulating varnishes and for laminating glass cloth.

Vulcanized silicone rubber is prepared in two principal forms: (1) as room-temperature-vulcanizing (RTV) elastomers, which are low-molecular-weight liquids that are cast or molded into desired shapes and then interlinked at room temperature, and (2) high-temperature-vulcanizing (HTV) elastomers, which are higher-molecular-weight gums that are mixed and processed like other elastomers. Silicone rubbers are usually strengthened by fillers such as silica; other fillers are mixed in to add bulk and color. Valued for their electrical-insulating properties, chemical stability, and the wide temperature range over which they retain resiliency, silicone rubbers are used mainly in O-rings, heat-resistant seals, caulks, gaskets, electrical insulators, flexible molds, and (owing to their chemical inertness).

Polyphosphazene

Polyphosphazenes are by far the largest class of inorganic polymers. Hundreds of different polymers of this type have been synthesized, with a range of physical and chemical properties that rival those of organic polymers. This class of polymers possesses many valued properties such as high thermal and chemical stability and comparatively low toxicity which makes them attractive materials for numerous biomedical, elastomeric, and electrochemical applications. The high thermal-oxidative stability of the polymer backbone has been explained by the high oxidation state of the main chain atoms which means that the thermodynamic driving force for oxidative skeletal cleavage is rather weak. Most of the other properties depend to a large extent on the type of substituents on the polymer backbone and on the chain architecture including glass-transition and melting temperatures, modulus and tensile strength as well as hydrolytic stability, solubility in solvents and compatibility with other polymers. These properties can be varied over a wide range by changing the type of side groups, the molecular weight, and cross-link density.

Some forms of unmodified and modified polyphosphazenes are depicted below. The polymer backbone of unmodified polyphosphazene consists of alternating phosphorus and nitrogen atoms, with two side groups, R, being attached to each phosphorus: (a). These side groups may be organic, organometallic, or inorganic. Also obtainable are polymers in which a portion of the phosphorus atoms has been replaced by sulfur (b) or carbon atoms (c) or in which polyphosphazenes are linked to other polymers such as polysiloxanes (d) or polyethers (e).

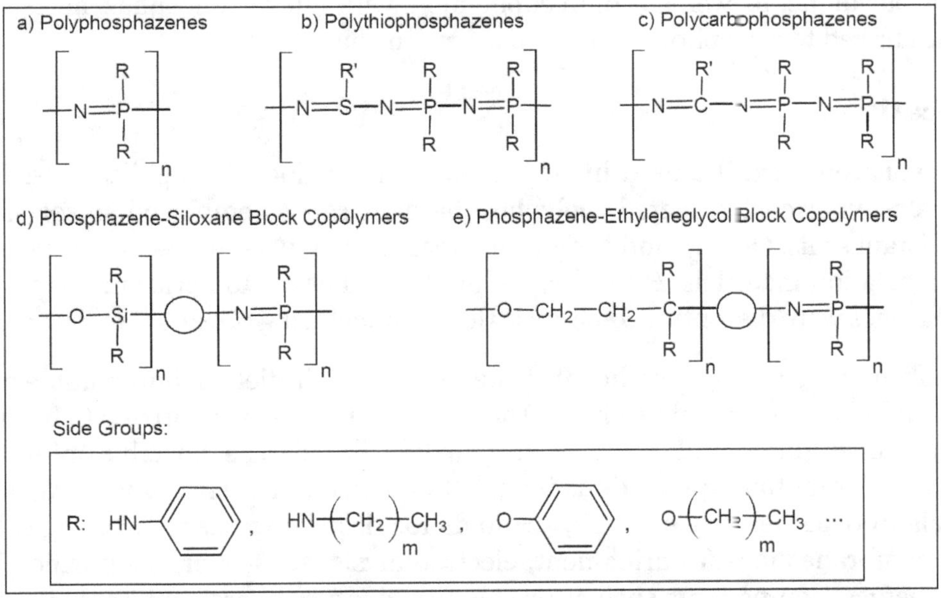

Synthesis

Polyphosphazenes are produced by a unique synthesis method that allows the types of side groups, R, to be varied over a very broad range. The most common method uses hexachlorocyclotriphosphazene (phosphonitrilic chloride trimer) as a starting material. It is first converted to poly(dichlorophosphazene) by ring-opening polymerization at elevated temperature (250°C) which yields polymers of high molecular weight (10^6 - 10^7 g/mol). The resulting linear polymers can be dissolved completely in organic solvents such as benzene, toluene, or tetrahydrofuran. Subsequent reaction of the dissolved linear poly(dichlorophosphazene) with sodium alkoxides or alkylene amines yields polyalkoxyphosphazenes and polyaminophosphazenes, respectively, whereas further heating of poly(dichlorophosphazene) causes cross-linking of the chains and yields insoluble inorganic elastomers.

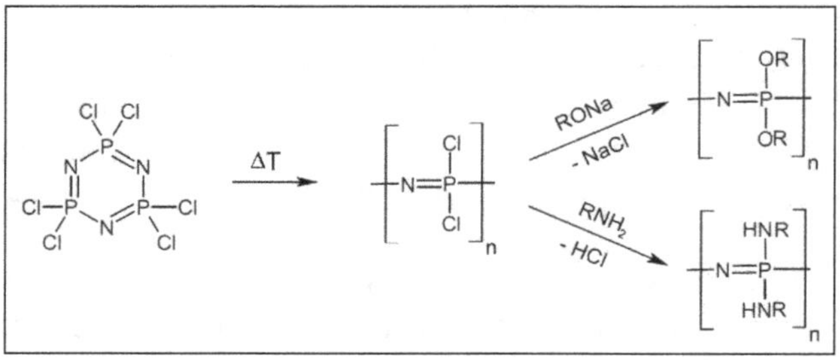

Besides linear polyphosphazenes, other molecular architectures such as block copolymers, comb copolymers, cyclo-linear polymers, and star polymers have also been synthesized. Particularly noteworthy are phosphazene-organic and phosphazene-polysiloxane block copolymers, copolymers with six-membered phosphazene rings in the polymer backbone, as well as comb copolymers with (short) phosphazene or organic branches linked to phosphorus atoms in the main chain.

Applications

Polyphosphazenes can be used in very demanding engineering applications. For example, elastomeric (fluorinated) polyphosphazenes can be employed as solvent- and heat resistant seals, O-rings and gaskets and foamed products as fire- and heat resistant sound-insulation materials. However, the production of elastomeric and foamed polyphosphazenes were discontinued because the manufacturing process was too expensive.

An important area where polyphosphazenes excel is in the field of implantable medical devices and pharmaceutical products. These polymers have been proven to be biocompatible, biodegradable, and bioactive, and can be tailored to be biostable or biodegradable. Major (potential) applications include biodegradable release and encapsulation materials in drug delivery systems; scaffolds for tissue engineering; biological separation membranes for air enrichment, electrodialysis, and water/gas purification; as well as medical components such as prosthetics, denture liners, and implants.

Polysilazane

Polysilazanes are a class of very heat stable inorganic polymers with a polymer backbone made up entirely of silicon-nitrogen bonds with either only hydrogen substituents attached to each silicon and nitrogen atom (perhydro-polysilazanes PHPS, -NH-SiH_2-)

or with additional organic substituents attached to each silicone (organo polysilazanes, OPSZ, -NH-SiR_2-). The pendent groups are typically hydrogen, alkyl, and/or aryl groups. Thus an almost unlimited number of different polymers belong to this class. However, only a small number if silazane monomer/polymers are commercially available.

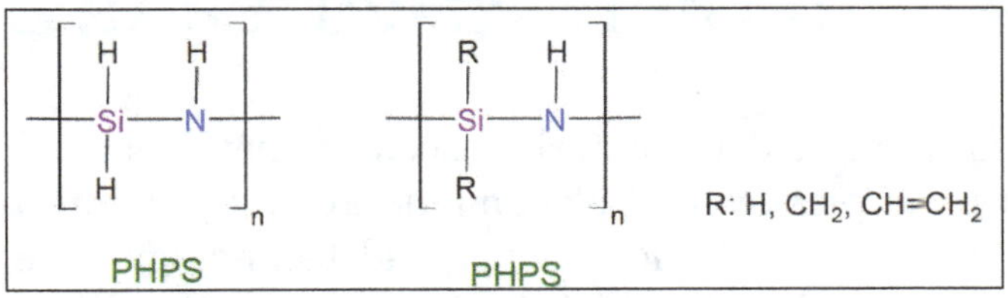

Polysilazanes are known for their high temperature and oxidative stability; excellent scratch, abrasion and impact resistance; high hardness (up to 5H for OPSZ and 9H for PHPS); and high resistance to weathering and many chemicals. Most or all of these properties are superior to those of polysiloxanes. Polysilazanes with nonpolar organic side groups have also outstanding non-stick and easy clean properties which are almost as as good as those of fluoropolymers (PTFE).

Polysilazanes are mainly used as precursors to SiO_2, Si_3N_4 and SiC ceramics. Pyrolysis / oxidation of these precursors provide a convenient method for preparing silicon-based ceramic products, such as fibers, coatings and fiber-reinforced ceramic-matrix composites (CMCs) which are often difficult to manufacture with conventional ceramic processing methods. However, the use of polysilazanes has been severely limited by the high price and by lack of availability. The most important applications appear to be protective and heat resistant coatings in the electronic, automotive, and aerospace industry (exhaust systems, engine parts, heat exchangers, ovens and electronic devices). Organo silazanes are also sometimes incorporated into one- and two-component coatings for car / train exteriors and architectural elements. These products protect smooth and non-absorbent surfaces against graffiti and dirt.

References

- Inorganic-polymer, chemical-engineering: sciencedirect.com, Retrieved 13, May 2020
- The-Properties-and-Latest-Application-of-Geopolymers-331191626: researchgate.net, Retrieved 26, August 2020
- Silicone, science: britannica.com, Retrieved 18, April 2020
- Polyphosphazenes, polymer-classes: polymerdatabase.com, Retrieved 07, January 2020
- Silazanes, polymer-brands: polymerdatabase.com, Retrieved 14, March 2020

5

Polymers and their Architecture

Polymer science is a subfield of materials science that involves polymers, mainly synthetic polymers like plastics and elastomers. Polymer architecture describes the shape of a single polymer molecule. Science and technology have undergone rapid developments in the past decade which has resulted in the discovery of significant tools and techniques in the field of polymer science and architecture which have been extensively detailed in this chapter.

Polymer Science

Polymer science is an interdisciplinary area comprised of chemical, physical, engineering, processing, and theoretical fields of science. Its objective is to provide the basis for the creation and characterization of polymeric materials and an understanding for the relationship between structure and property.

Future Scope of Polymer Science

The main concerns for humans in the future will be energy & resources, food, health, mobility & infrastructure and communication. There is no doubt that polymers will play a key role in finding successful ways in handling these challenges. Polymers will be the material of the new millennium and the production of polymeric parts i.e. green, sustainable, energy-efficient, high quality, low-priced, etc. will assure the accessibility of the finest solutions round the globe. Synthetic polymers have since a long time played a relatively important role in present-day medicinal practice. Many devices in medicine and even some artificial organs are constructed with success from synthetic polymers. It is possible that synthetic polymers may play an important role in future pharmacy, too. Polymer science can be applied to save energy and improve renewable energy technologies.

Biopolymers could especially increment as more solid adaptations are produced, and the cost to fabricate these bio-plastics keeps ongoing fall. Bio-plastics can supplant routine plastics in the field of their applications likewise and can be utilized as a part of various areas, for example, sustenance bundling, plastic plates, mugs, cutlery, plastic stockpiling packs and in this way can help in making environment economical.

In areas of applications of plastics materials, a well-known long standing example is electrical industries have led to increasing acceptance of plastics for plugs, sockets, wire and cable insulations and for housing electrical and electronic equipment. The major polymer targeting industries of the present day life includes Ceramic industries, in stem cell biology and Regenerative Medicine, packaging industries, in retorting method used for food processing industries in automotive industries, in aerospace industries and in electrical and electronic industries.

- Polymers in Stem Cell Biology
- Self-Healing and Reprocess-able Polymer Systems
- Smart Polymers
- Green Synthesis of Functional Materials
- In Gene Delivery Systems
- Ceramic Industry
- Biopolymers in Drug Delivery
- Market growth of Polymers

Polymer Architecture

Polymers occupy a major portion of materials used for controlled release formulations and drug-targeting systems because this class of materials presents seemingly endless diversity in topology and chemistry. This is a crucial advantage over other classes of materials to meet the ever-increasing requirements of new designs of drug delivery formulations. The polymer architecture (topology) describes the shape of a single polymer molecule. Every natural, semi natural, and synthetic polymer falls into one of categorized architectures: linear, graft, branched, cross-linked, block, star-shaped, and dendron/ dendrimer topology. Although this topic spans a truly broad area in polymer science.

The basic structure of a polymer is a chain. The chain is formed by the backbone of the polymer. It may be decorated with pendant groups, such as methyls in polypropylene, chlorines in polyvinyl chloride or phenyls in polystyrene, but the basic structure is like a long string.

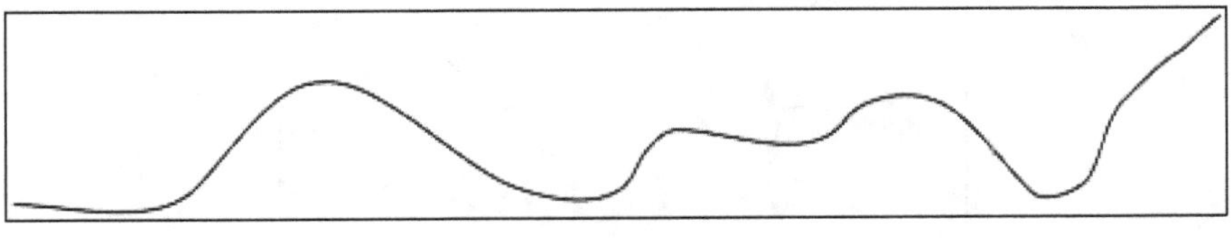

Sometimes, the architecture of a polymer can be more complex than that. For example, low-density polyethylene has branches coming off the main chain. Consequently, it is described as a branched polymer.

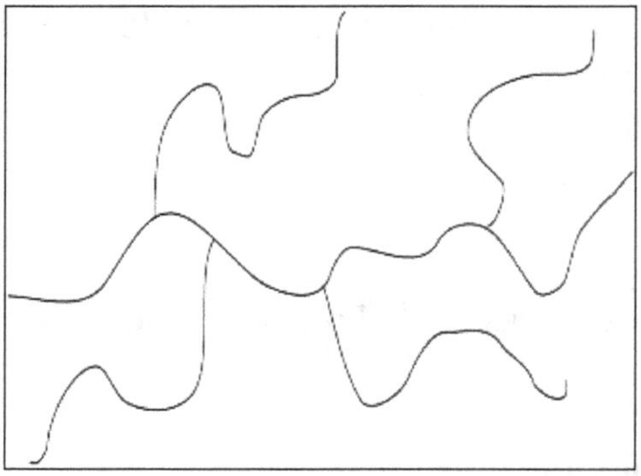

In some cases, branching is so extensive that the polymer does not resemble a chain at all. Some of these highly-branched polymers are called dendrimers. Dendrimers are tree-like structures that branch out in all directions from a central point. Viewed from a distance, a dendrimer would have an overall shape that is more like a fuzzy ball than like a coiled chain.

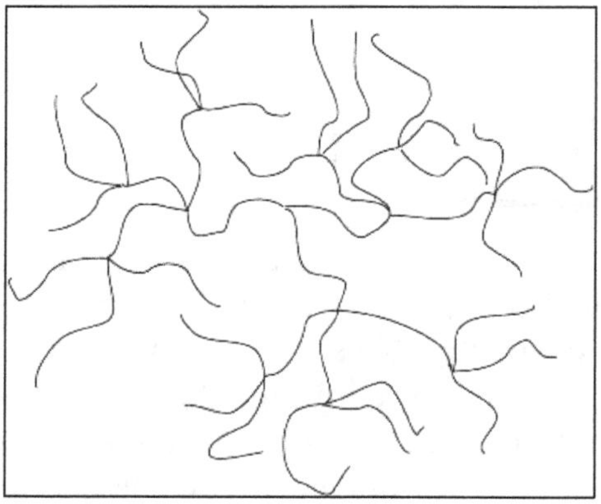

If the branches on one branched chain connect directly to other chains, then the polymer is said to be cross linked. Cross linked polymers can have extra strength and rigidity because they are less flexible than polymers in which the chains are able to move independently. However, just the right amount of crosslinking can make a polymer more elastic, making it bounce back to its original form after it is flexed.

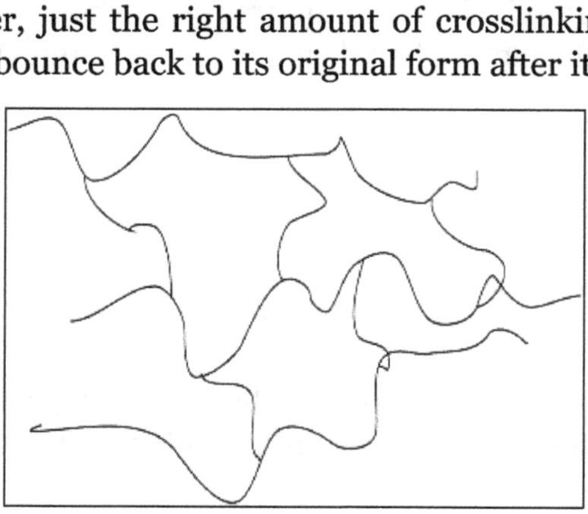

Crosslinking can actually be of two different types: "chemical" or "physical". Chemical crosslinks are permanent covalent bonds between the chains. Physical crosslinks are more temporary; they are formed by strong intermolecular attractions, such as hydrogen bonds, between chains.

Copolymers

The structure of polymers can vary in how their chains are arranged. Additional structural complexity arises if the polymer is a copolymer. Regular polymers are made from one kind of monomer. Copolymers are formed from two different monomers.

If there are two different monomers, they may be arranged in different ways along the chain. Maybe the two monomers simply alternate, one after the other: that's called an alternating copolymer. Maybe they are distributed at random: that's a random or statistical copolymer. Maybe they are clustered in two separate groups: that's called a block copolymer.

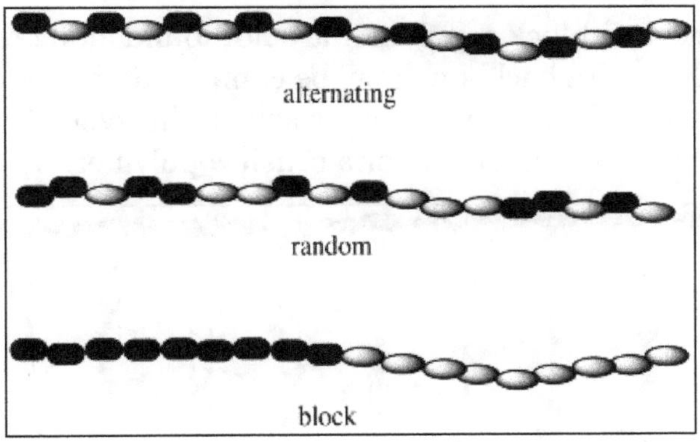

These copolymers are different from polymer blends. Polymer blends are just mixtures of two different kinds of polymers. Both polymer blends and copolymers are important. Either one might allow different properties of each polymer, such as strength and flexibility, to be brought together in one material.

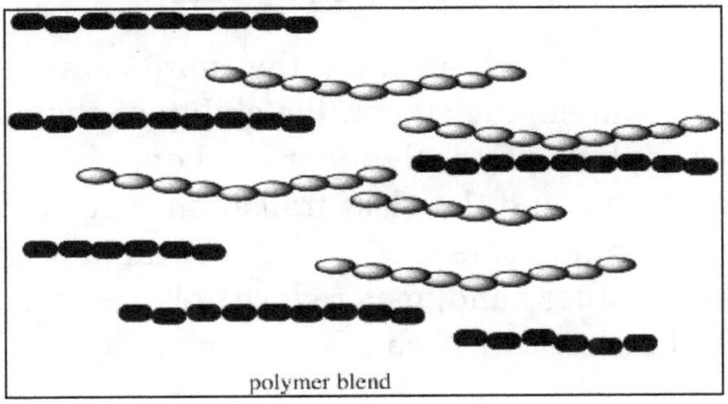

There are different kinds of block copolymers, depending on how the blocks of each monomer are arranged in the chain. The simplest kind of chain would contain one long block of the first monomer followed by one long block of the other. That's called a diblock (or AB diblock) copolymer . If two blocks of the same polymer are found on the ends, with a different one in the middle, that would be called an ABA triblock copolymer. If there were three different blocks from three different monomers, we would have an ABC triblock copolymer instead. There are lots of possible variations.

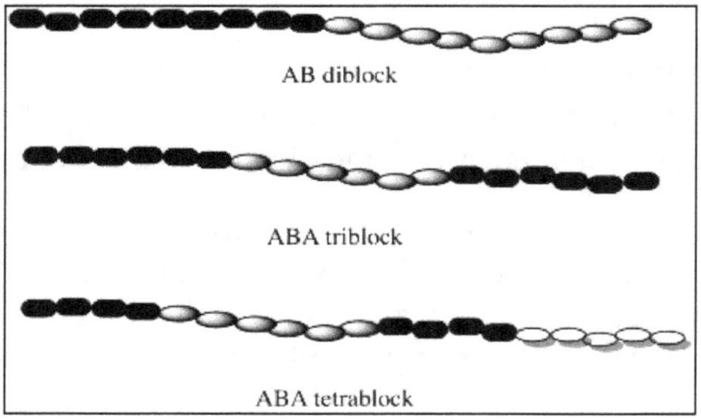

Sometimes, the blocks in a block copolymer are not found along the chain. For example, in some cases, the entire backbone may be composed of one monomer, with pendant chains attached to the backbone. A material with this type of architecture is called a graft copolymer, or sometimes a comb or a brush copolymer.

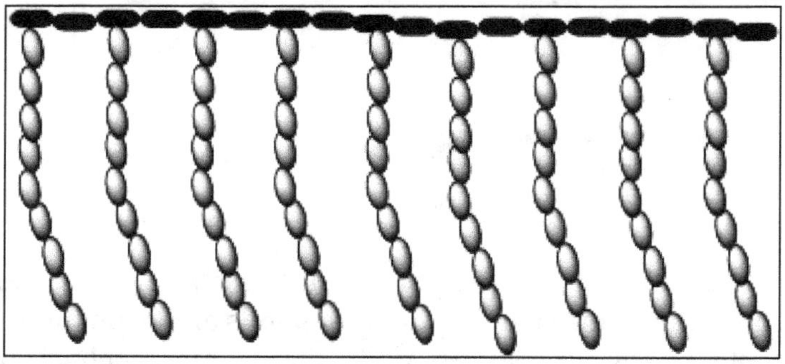

Effect of Architecture on Physical Properties

In general, the higher degree of branching, the more compact a polymer chain is. Branching also affects chain entanglement, the ability of chains to slide past one another, in turn affecting the bulk physical properties. Long chain branches may increase polymer strength, toughness, and the glass transition temperature (T_g) due to an increase in the number of entanglements per chain. A random and short chain length between branches, on the other hand, may reduce polymer strength due to disruption of the chains' ability to interact with each other or crystallize.

An example of the effect of branching on physical properties can be found in polyethylene. High-density polyethylene (HDPE) has a very low degree of branching, is relatively stiff, and is used in applications such as bullet-proof vests. Low-density polyethylene (LDPE), on the other hand, has significant numbers of both long and short branches, is relatively flexible, and is used in applications such as plastic films.

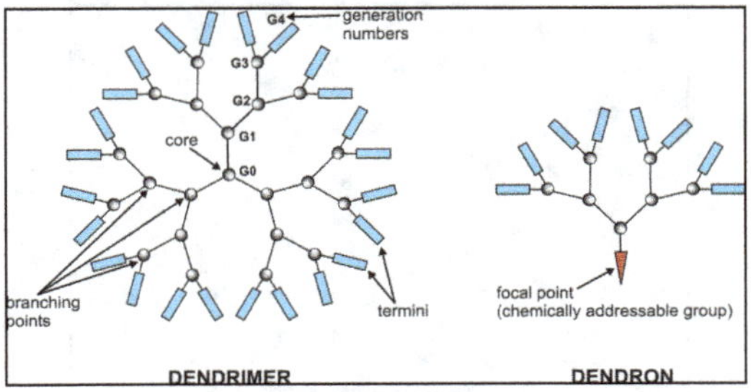

Dendrimer and dendron.

Dendrimers are a special case of branched polymer where every monomer unit is also a branch point. This tends to reduce intermolecular chain entanglement and crystallization. A related architecture, the dendritic polymer, are not perfectly branched but share similar properties to dendrimers due to their high degree of branching.

The degree of branching that occurs during polymerisation can be influenced by the functionality of the monomers that are used. For example, in a free radical polymerisation of styrene, addition of divinylbenzene, which has a functionality of 2, will result in the formation of branched polymer.

Branched Polymers

The properties of polymers are strongly affected by their molecular weight and molecular weight distribution as well as by their chain architecture, particularly by the amount of branching. The effect of branching on the polymer properties depends on the number and length of the branches. Short branches interfere with the formation of crystals, that is, they reduce the amount of crystallinity whereas long branches undergo side chain crystallization because they are able to form lamellar crystals of their own. In the case of polyethylene, significant side chain crystallization can be expected around 40 carbon atoms. Long side chains also have a noticeable effect on the flow properties of the polymer, particularly when the length of the branches exceeds the average critical entanglement length. In that case, even a small amount of branching will greatly affect the processing properties.

Branching plays an important role in the performance of polyolefins. For example, linear polyethylene has a high degree of crystallinity and rather poor mechanical properties. Even a small amount of long-chain branches can significantly improve the mechanical properties and the processability of a polyolefin. This is particularly true when the polyolefin has a narrow molecular weight distribution and a high degree of crystallinity. Both the degree of branching as well as the length of the branches affects the density which can vary considerably. Typically, the higher the density of the polymer the higher the degree of crystallinity and the stiffer, harder, and stronger the polymer.

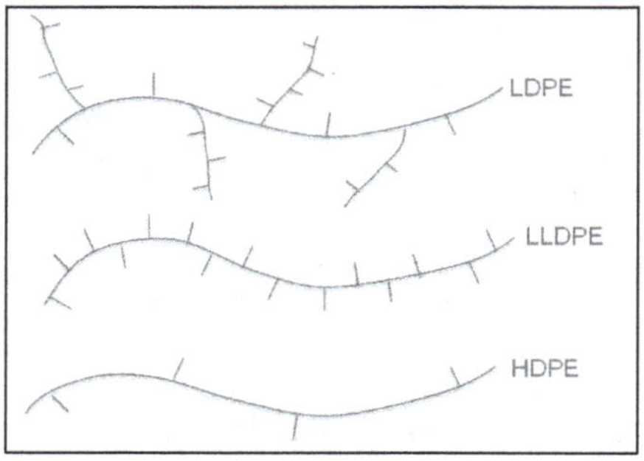

Today, many grades of polyethylene are produced with single-site metallocene catalysts. These polymers are mostly linear and have a high degree of crystallinity. However, these polymers are rather brittle. To increase the flexibility, ethlyene is typically copolymerized with low molecular weight alkenes such as propene, butene-1, hexene-1, 4-methyl-pentene-1 or octene-2 which introduce short chain branches on the mostly linear polymer chain. In general, the toughness and stress crack resistance increases with increasing chain length of the branches. Thus, copolymerization of ethylene with hexene-1 produces tougher polymers than copolymerization with propylene.

The molecular weight distribution and degree and number of chain branching also affect the melt flow properties such as shear thinning behavior and melt flow index. Because of significant differences in molecular structure between highly branched low density polyethylene (LDPE) and linear low and high density polyethylene (LLDPE, HDPE), in particular in molecular weight and chain branching, there are significant differences in the rheological behavior between these types of polyolefins. The effect of long-chain branching on the viscosity was analyzed by Bueche. He found following relationship:

$$\eta_b / \eta_l = g^{3.5}$$

where η_b is the viscosity of the branched polymer and η_l that of a chemically similar linear polymer of same molecular weight. The parameter g is the ratio of the mean square radii of the branched and linear polymer. Since g is less than unity, Bueche's relation predicts a lower viscosity for long-branched polymers.

Branching in Radical Polymerization

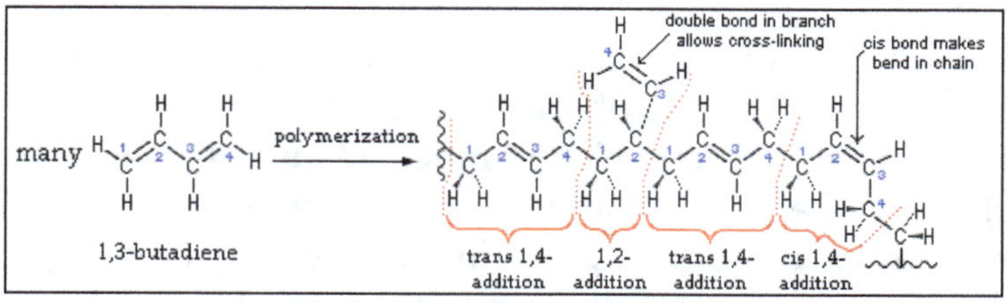

Polymerization of 1,3-butadiene.

In free radical polymerization, branching occurs when a chain curls back and bonds to an earlier part of the chain. When this curl breaks, it leaves small chains sprouting from the main carbon backbone. Branched carbon chains cannot line up as close to each other as unbranched chains can. This causes less contact between atoms of different chains, and fewer opportunities for induced or permanent dipoles to occur. A low density results from the chains being further apart. Lower melting points and tensile strengths are evident, because the intermolecular bonds are weaker and require less energy to break.

The problem of branching occurs during propagation, when a chain curls back on itself and breaks - leaving irregular chains sprouting from the main carbon backbone. Branching makes the polymers less dense and results in low tensile strength and melting points. Developed by Karl Zieglerand Giulio Natta in the 1950s, Ziegler-Natta catalysts (triethylaluminium in the presence of a metal(IV) chloride) largely solved this problem. Instead of a free radical reaction, the initial ethene monomer inserts between the aluminium atom and one of the ethyl groups in the catalyst. The polymer is then able to grow out from the aluminium atom and results in almost totally unbranched chains. With the new catalysts, the tacticity of the polypropene chain, the alignment of alkyl groups, was also able to be controlled. Different metal chlorides allowed the selective production of each form i.e., syndiotactic, isotactic and atactic polymer chains could be selectively created.

However, there were further complications to be solved. If the Ziegler-Natta catalyst was poisoned or damaged then the chain stopped growing. Also, Ziegler-Natta monomers have to be small, and it was still impossible to control the molecular mass of the polymer chains. Again new catalysts, the metallocenes, were developed to tackle these problems. Due to their structure they have less premature chain termination and branching.

Branching Index

The branching index measures the effect of long-chain branches on the size of a macromolecule in solution. It is defined as g = <sb2>/<sl2>, where sb is the mean square radius of gyration of the branched macromolecule in a given solvent, and sl is the mean square radius of gyration of an otherwise identical linear macromolecule in the same solvent at the same temperature. A value greater than 1 indicates an increased radius of gyration due to branching.

Star Polymer

Star polymers consist of several linear polymer chains connected at one point. Star molecules prepared by anionic polymerization had been examined prior to the development of CRP. However due to the limitations of ionic polymerization the composition and functionality of the materials were limited but their compact structure and globular shape provide them with a set of unique properties, such as low solution viscosity, and the core shell architecture enabled several potential applications spanning a range from thermoplastic elastomer's to drug carriers.

Based on the chemical compositions of the arm species, star polymers can be classified into two categories: homo-arm (or regular) star polymers or mikto-arm (or heteroarm) star copolymers. Homo-arm star polymers consist of a symmetrical structure comprising radiating arms with identical chemical composition and similar molecular weight. In contrast, a miktoarm star molecule contains two or more arm species with different chemical compositions and/or molecular weights and/or different peripheral functionality. There are several approaches that can be employed for synthesis of star copolymers.

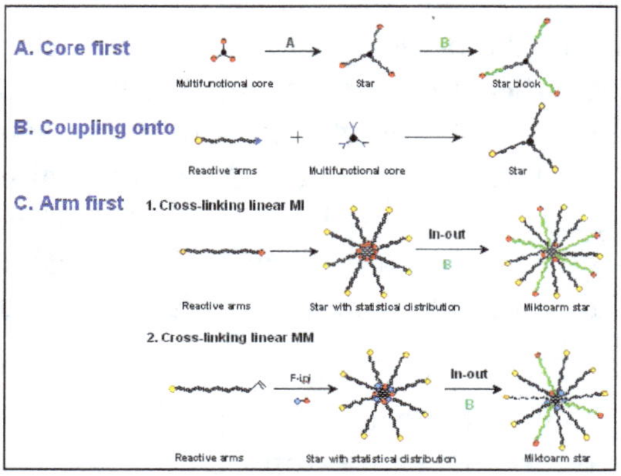

As suggested above, star polymers can be synthesized by variations on one of three methods:

- The *"core-first"* approach, where the controlled polymerization is conducted from either a well defined initiator with a known number of initiating groups or a less well defined multifunctional macromolecule.
- An approach that until recently had not received as much attention is the *"coupling onto"*approach where a tele-functional linear molecule is reacted with a preformed core molecule containing complementary functionality. In order to improve the coupling efficiency, a highly efficient organic coupling reaction required, such as click reactions. The preparation of stars using click chemistry could be applied to almost any of the strategies discussed on this page.
- There are two approaches to the *"arm-first"* synthesis of star polymers. One is where a linear "living" copolymer chain, or added macroinitiator, is linked by continuing the copolymerization of the mono-functional macroinitiator with a divinyl monomer forming a cross linked core.
- The other, is the direct copolymerization of a macromonomer with a divinyl monomer in the presence of a low molecular weight initiator.

- A combination of "arm-first" and "core-first" methods is particularly useful for synthesis of miktoarm star copolymers. One employs the retained initiating functionality in the formed "arm first" core to initiate the polymerization of a second monomer in a "grafting out" or a "grafting from" copolymerization.

Core First

The following section on nano-composites formed by grafting from the surface of a functionalized particle is an evolution of the "core first" approach to synthesis of star macromolecules. Initially a low molecular weight multi-functional molecule was used in a "grafting from" reaction to form star macromolecules with a well defined number of arms. Use of hexakis(chloromethyl) benzene as a well defined multifunctional initiator for the polymerization of styrene provided the first multi-arm polymer prepared using ATRP. The composition of the core and arms were quickly expanded.

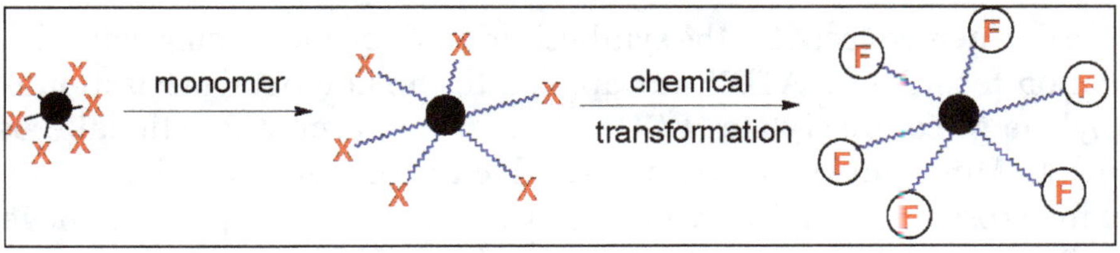

Since the tethered chains in a grafting from reaction retain their terminal functionality they could be chain extended to form star block copolymers and/or the radically transferable atoms on the chain ends (X) could be converted to other functional groups suitable for post-polymerization functionalization reactions.

Less well-defined multi arm star structures were prepared by grafting from a soluble multifunctional hyperbranched core. The core was prepared by polymerizing a molecule containing both a reactive group suitable to initiate a CRP and a polymerizable double bond, an initiator/monomer or inimer. Self-condensing vinyl polymerization using ATRP has been applied to inimers, i.e. monomers that also contain activated halogen atoms, such as α-bromoesters and benzyl halides. For example, when 2-bromopropionyloxyethyl acrylate was polymerized in the presence of a copper catalyst with 4,4'-dinonyl-2,2'-bipyridine ligands, a hyperbranched polymer with degree of branching ~0.5 and *DP* = 78 was obtained. Each soluble hyperbranched polymer contained on average 78 active bromopropionyloxy groups which were used to initiate the polymerization of methyl acrylate forming a star with 78 arms.

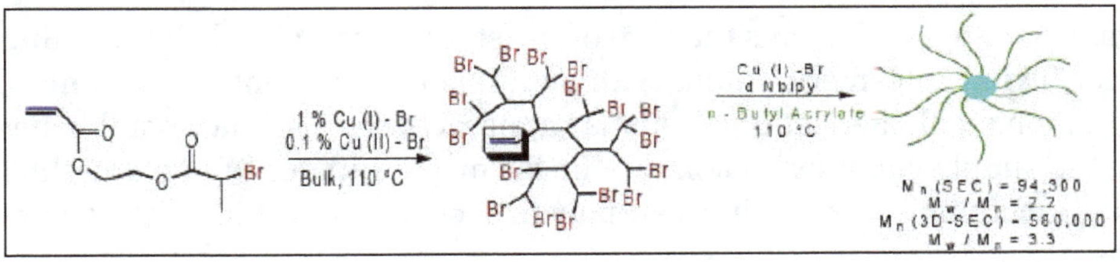

A simple sequential polymerization of a crosslinker followed by polymerization of a monomer provides a broadly applicable approach to star copolymers. This novel method, termed “star from in situ generated core”, belongs to the broader category of “core-first” methodology and presents an alternative strategy for star synthesis when compared to the traditional “arm-first” method, in which monomer is polymerized first followed by formation of the core by (co)polymerization of a cross-linker.

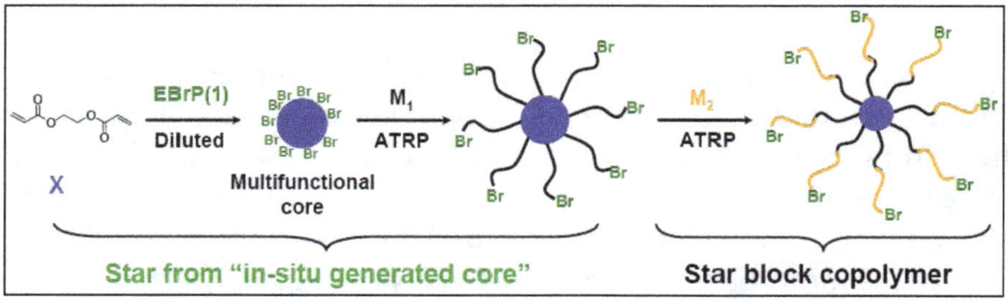

To illustrate this new concept for the synthesis of star polymers using controlled radical polymerization techniques, ATRP was applied to the homopolymerization of a cross linker, ethylene glycol diacrylate (EGDA) to generate a multifunctional cross linked core (nanogel). Monovinyl monomers were added to the reaction mixture at high conversion of the cross linker and were polymerized from the poly EGDA nanogel macroinitiator (MI) to form the star arms. A spectrum of different acrylate monomers were polymerized from the first formed core providing star polymers with different compositions for the arms.

Several parameters affected the structure of the star, such as the initial concentration of the cross linker, the molar ratio of cross linker to initiator, and the moment of addition of the monovinyl monomer to the ongoing polymerization reaction. The star polymers preserved the initiating sites at the chain ends, and they were further used as star MIs for arm extension by polymerization of a second monovinyl monomer to form a star block copolymer.

Indeed this approach of preparing well defined polymers by copolymerization of a cross linker and a monomer can be expanded to provide access to a full range of stars/gels/networks.

Arm First

Macroinitiator Approach

The “arm first” approach forms the core of the star macromolecule by coupling monofunctional “living” polymeric chains with a difunctional reagent and was first applied to living anionic polymerization. A similar approach was also successful using ATRP. Initially the simple chain extension of a linear macroinitiator with a crosslinker provided star macromolecules with broad polydispersity as a result of star-star coupling reactions.

There are several parameters in an ATRP that should be carefully controlled in order to maximize the yield of stars and prevent/reduce star-star coupling reactions. Some detailed studies have been carried out on the coupling of monofunctional polystyrenes and polyacrylates with divinylbenzene (DVB) and di(meth)acrylates to prepare star polymers and the following guidelines were developed:

- The ratio of di-functional reagent to growing chains seems to be optimal in the range of 10-20.
- Monomer conversion (or reaction time) has to be carefully controlled and stopped before star-star coupling occurs. It seems that ~5% of arms can not be incorporated into the star macromolecules under typical one step conditions.
- Higher yields of stars are observed for polyacrylates than for polystyrenes. This may be attributed to a higher proportion of terminated chains (due to slower propagation and higher concentration of radicals) in styrene polymerization under "standard" ATRP conditions. (The use of ARGET ATRP for the synthesis of the active growing polymer chains should change both these observations.)
- The choice of the diffunctional reagent is important and reactivity should be similar to, or lower than that of the arm-building monomers.
- Halogen exchange slightly improves the efficiency of star formation.
- Solvent, temperature and catalyst concentration should be also optimized.

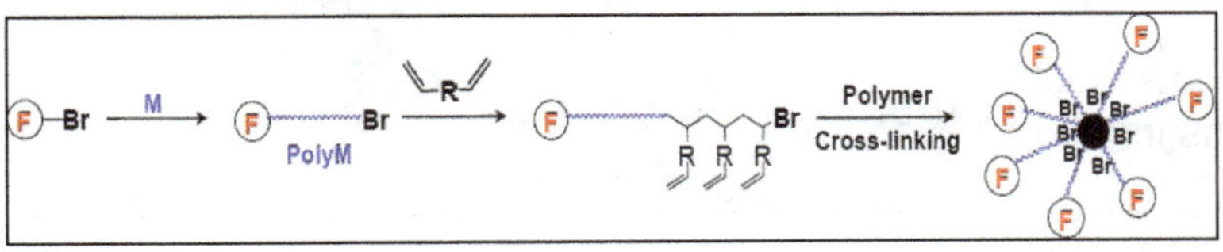

Under typical conditions, ~50 arms are coupled into star structures. Apparent molecular weights, measured by GPC, show a 10-20-fold increase of molecular weight (from M_n = 10,000 to M_n ~ 150,000) but light scattering indicates that the stars have a much higher molecular weight (M_n~500,000).

A range of initiators, monomers and cross linkers was examined for the preparation of star molecules with peripheral functionality and cores of differing phobicity.

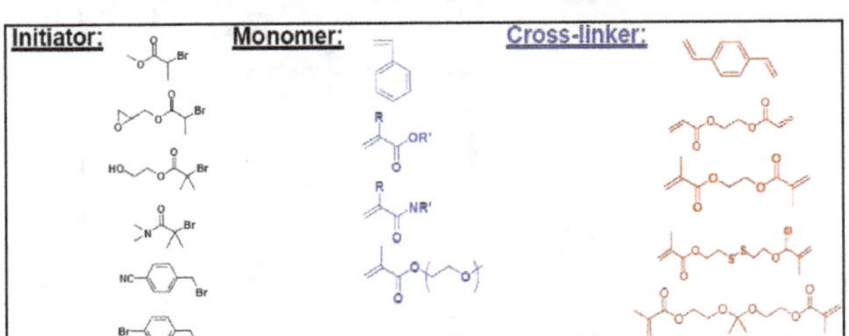

Another approach to form star polymers in high yield was via cross-linking self organized amphiphilic macroinitiators by AGET ATRP in aqueous dispersed media. Linear poly(ethylene oxide)-b-polystyrene (PEO-PS-Cl) block copolymers were the arm precursors. The amphiphilic block copolymers with halogen chain-end functionality formed divinyl cross-linker swollen micelles in water and then were cross-linked by the cross-linkers when polymerization was initiated. Due to the formation of micelles before the polymerization was initiated, star-star or star-linear chain reactions were not required for the star formation. This suppressed star-star coupling reactions and resulted in the formation of star polymers with low PDI (Mw/Mn < 1.1) but still high molecular weight (over 1000 kg/mol).

Macromonomer Approach

Another approach to arm first star copolymers, with available core containing initiating functionality, is the CRP of higher molecular weight macromonomers in a pure homo-polymerization initiated with a small molecule initiator which usually leads to a "brush molecule" with a degree of polymerization of 10-25, which from a topological standpoint can be considered a star.

An alternate approach is a copolymerization of macromonomers with a divinyl cross-linking monomer.

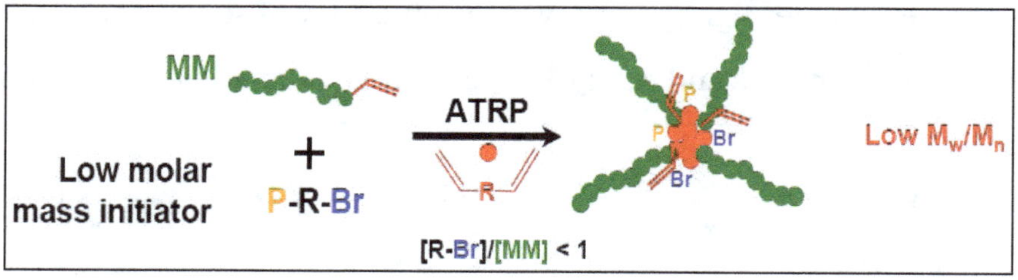

The sequential addition of additional initiator and crosslinker to the reaction increases the number of macromonomer units incorporated into each star.

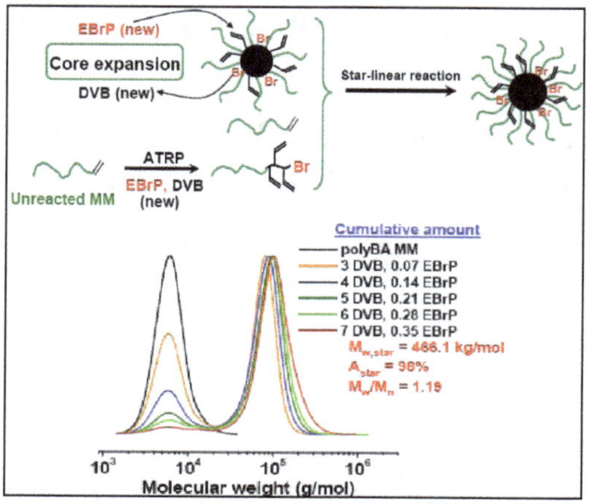

A potential application for homo-arm stars was illustrated by preparation of two well defined star macromolecules with two oppositely charged arm structures; a poly[2-(dimethylamino)ethyl methacrylate] (PDMAEMA) star and a poly(acrylic acid) (PAA) star with cross-linked cores. Exploitation of the electrostatic interactions between the polyelectrolyte arms of PDMAEMA star and PAA star polymers generated alternating polyelectrolyte multilayer LbL films using layer-by-layer (LbL) assembly.

Mikto-arm Star Copolymers

In-out Method

Both approaches to form star molecules using the arm first approach retain the initiating functionality within the core of the star and therefore provide a simple approach to form mikto-arm stars by conducting a controlled polymerization of a second monomer from the accessible initiator functionality in the core.

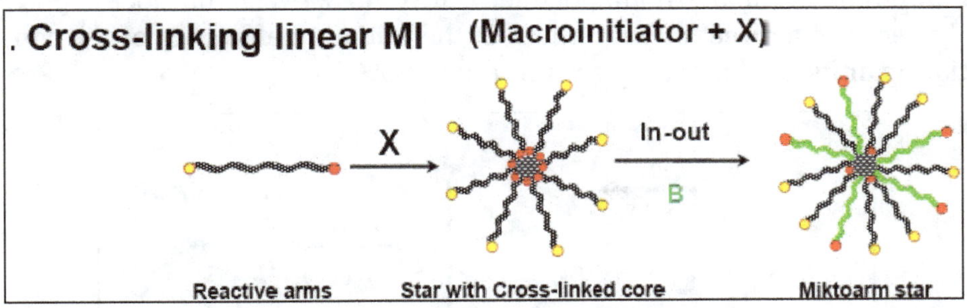

As noted above one arm-first procedure involves the synthesis of a linear mono-functional polymer chain polyA, which is used as a macroinitiator (MI) in a subsequent cross-linking reaction using a divinyl compound to produce a $(polyA)_n$-polyX star polymer, where polyX represents the core of the star polymer and n is the number of polyA arms. The initiating sites are preserved within the core of the star polymer (i.e., alkyl halide groups in ATRP) and the $(polyA)_n$-polyX star polymer can be used as a multifunctional star initiator in a chain extension reaction with a different monomer, B, to yield a miktoarm star copolymer, $(polyA)_n$-polyX-$(polyB)_m$. This combination method for synthesizing miktoarm star copolymers was termed the "in-out" method.

The efficiency of initiation of the second set of arms is dependent on the compactness of the first formed core with less densely crosslinked cores providing more efficient initiation i.e., a greater fraction of the encapsulated initiator sites, for the grafting out polymerization.

Normally, one seeks to, or has to, form a chemically stable core. However, it is possible to select a crosslinking agent with a degradable link between the two functional crosslinking groups and prepare a material with a degradable core.

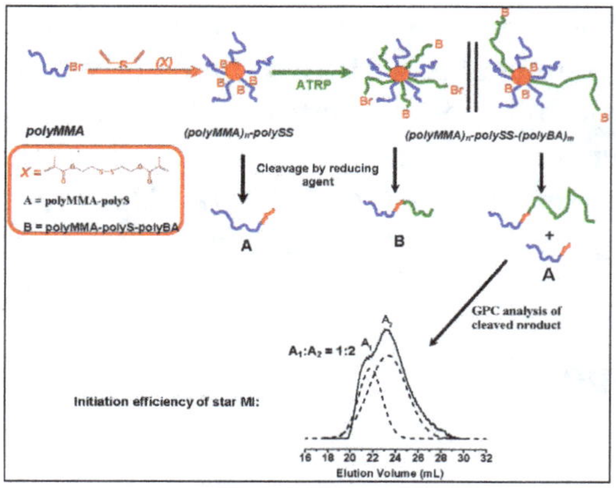

This was accomplished by linking the first formed arms with a dimethacrylate crosslinker containing a disulfide link between the methacrylate units. The mikto-arm star copolymer could be degraded in a reducing environment to form a mixture of an AB block copolymer and some residual A-homopolymer. The ratio between the block copolymer and homopolymer gave a direct measurement of the initiating efficiency of the constrained core initiating units; in this example it was only ~20%.

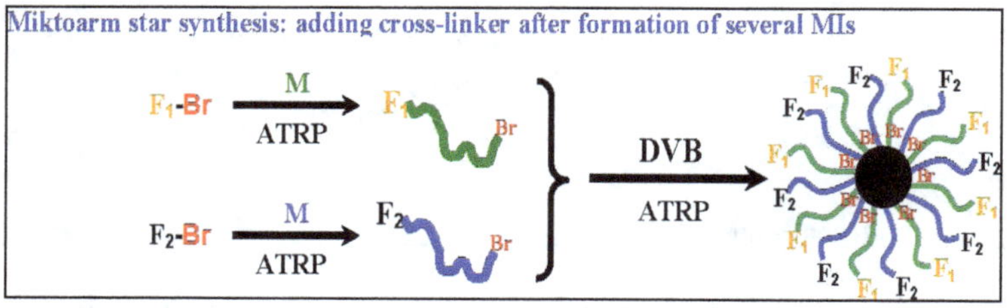

Liquid chromatography under the critical conditions (LCCC) for each of the homopolymers arms provided proof that the star structure a miktoarm star copolymer containing two or more compositionally different arm species in one molecule, and was not a mixture of different homo-arm star polymers. Miktoarm star copolymers containing five kinds of arms were synthesized for the first time by copolymerizing a mixture of five linear MIs with different chemical composition. When linear MMs were partially or completely used as arm precursors instead of MIs, miktoarm star copolymers with high star yield and low polydispersity were successfully synthesized.

A recent example of a combination of an "arm first" and a "core first" approach is "grafting onto" a functionalized core via "click" chemistry. Three-arm and four-arm star polystyrene polymers were synthesized by a combination of ATRP and click coupling chemistry. The click reaction between an azido-terminated polystyrene (PS-N3) and an alkyne-containing multifunctional compound proved to be fast and efficient. When an azido-terminated polystyrene polymer was reacted with a trialkyne-containing or tetraalkyne-containing compound, the yields of 3-arm star and 4-arm star polymers were around 90% and 83%, respectively.

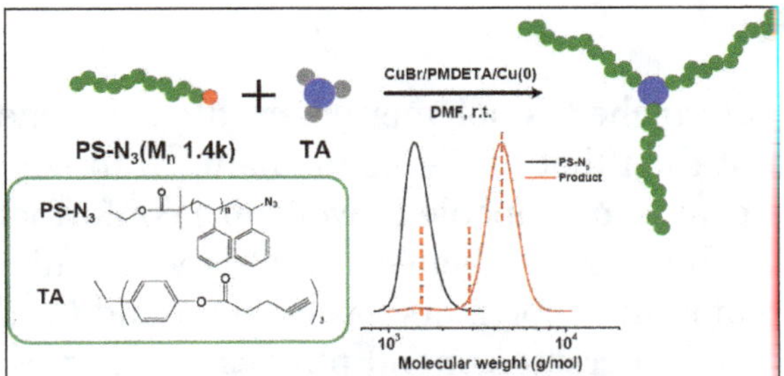

	M_n PS-N_3	PS-N_3/alkyne	Area fraction in GPC		
			3-arm star	PS-PS	PS
1	1,400	1/1	0.90	0.08	0.02
2	1,400	1/1.1	0.69	0.25	0.07
3	6,800	1/1	0.83	0.12	0.05

The influence of several parameters on the efficiency of the click coupling reaction was studied, including the molecular weight of the PS-N3 polymer, the presence of an added reducing agent, Cu°, and the stoichiometry between the azido and alkynyl groups. The results indicated that the yield of the coupled product was higher when a lower molecular weight PS-N3 was employed in conjunction with a small amount of reducing agent, and the molar ratio of azido and alkynyl groups was close to 1.

Graf Polymer

Graft polymers are segmented copolymers with a linear backbone of one composite and randomly distributed branches of another composite. The picture labeled "graft polymer" shows how grafted chains of species B are covalently bonded to polymer species A. Although the side chains are structurally distinct from the main chain, the individual grafted chains may be homopolymers or copolymers. Graft polymers have been synthesized for many decades and are especially used as impact resistant materials, thermoplastic elastomers, compatibilizers, or emulsifiers for the preparation of stable blends or alloys. One of the more well known examples of a graft polymer is high impact polystyrene, which consists of a polystyrene backbone with polybutadiene grafted chains.

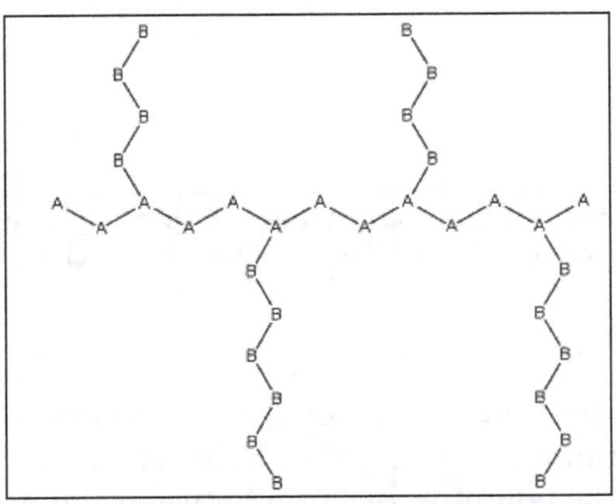

The graft copolymer consists of a main polymer chain or backbone (A) covalently bonded to one or more side chains (B).

General Properties

Graft copolymers are a branched copolymer where the components of the side chain are structurally different than that of the main chain. Graft copolymers containing a larger quantity of side chains are capable of wormlike conformation, compact molecular dimension, and notable chain end effects due to their confined and tight fit structures. The preparation of graft copolymers has been around for decades. All synthesis methods can be employed to create general physical properties of graft copolymers. They can be used for materials that are impact resistant, and are often used as thermoplastics elastomers, compatibilizers or emulsifiers for the preparation of stable blends or alloys. Generally, grafting methods for copolymer synthesis results in materials that are more thermostable than their homopolymer counterparts. There are three methods of synthesis, grafting to, grafting from, and grafting through, that are used to construct a graft polymer.

Synthesis Methods

There are many different approaches to synthesizing graft copolymers. Usually they employ familiar polymerization techniques that are commonly used such as atom transfer radical polymerization (ATRP), ring-opening metathesis polymerization (ROMP), anionic and cationic polymerizations, and free radical living polymerization. Some other less common polymerization include radiation-induced polymerization, ring-opening olefin metathesis polymerization, polycondensation reactions, and iniferter-induced polymerization.

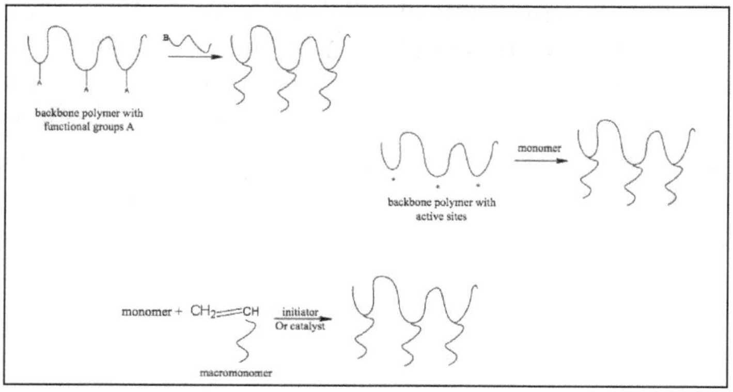

The three common methods of synthesis: grafting to (top left), grafting from (middle right), grafting through (bottom left), and their generalized reaction scheme are featured.

Grafting

The grafting to method involves the use of a backbone chain with functional groups A that are distributed randomly along the chain. The formation of the graft copolymer originates from the coupling reaction between the functional backbone and the end-groups of the branches that are reactive. These coupling reactions are made.

possible by modifying the backbone chemically. Common reaction mechanisms used to synthesize these copolymers include free-radical polymerization, anionic polymerization, atom-transfer radical-polymerization, and living polymerization techniques.

Copolymers that are prepared with the grafting-to method often utilize anionic polymerization techniques. This method uses a coupling reaction of the electrophilic groups of the backbone polymer and the propagation site of an anionic living polymer. This method would not be possible without the generation of a backbone polymer that has reactive groups. This method has become more popular with the rise of click chemistry. A high yield chemical reaction called atom transfer nitroxide radical coupling chemistry is for the grafting-to method for polymerization.

Grafting from

In the grafting-from method, the macromolecular backbone is chemically modified in order to introduce active sites capable of initiating functionality. The initiating sites can be incorporated by copolymerization, can be incorporated in a post-polymerization reaction, or can already be a part of the polymer. If the number of active sites along the backbone participates in the formation of one branch, then the number of chains grafted to the macromolecule can be controlled by the number of active sites. Even though the number of grafted chains can be controlled, there may be a difference in the lengths of each grafted chain due to kinetic and steric hindrance effects.

Grafting from reactions have been conducted from polyethylene, polyvinylchloride, and polyisobutylene. Different techniques such as anionic grafting, cationic grafting, atom-transfer radical polymerization, and free-radical polymerization have been used in the synthesis of grafting from copolymers.

Graft copolymers that are employed with the grafting-from method are often synthesized with ATRP reactions and anionic and cationic grafting techniques.

Grafting Through

The grafting through, also known as the macromonomer method, is one of the simpler ways of synthesizing a graft polymer with well defined side chains. Typically a monomer of a lower molecular weight is copolymerized with free radicals with an acrylate functionalized macromonomer. The ratio of monomer to macromonomer molar concentrations as well as their copolymerization behavior determines the number of chains that are grafted. As the reaction proceeds, the concentrations of monomer to macromonomer change causing random placement of branches and formation of graft copolymers with different number of branches. This method allows for branches to be added heterogeneously or homogeneously based on the reactivity ratio of the terminal functional group on the macromolecular to the monomer. The difference in distribution of grafts has significant effects on the physical properties of the grafted copolymer. Polyethylene, polysiloxanes and poly(ethylene oxide) are all macromonomers that have been incorporated in a polystyrene or poly(methyl acrylate) backbone.

The macromonomer (grafting through) method can be employed using any known polymerization technique. Living polymerizations give special control over the molecular weight, molecular weight distribution, and chain-end functionalization.

Applications

Graft copolymers became widely studied due to their increased number of applications due to their unique structures relative to other copolymers such as alternating, periodic, statistical, and block copolymers which generally linear chains.

Some common applications of graft copolymers include:

- Membranes for the separation of gases or liquids
- Hydrogels
- Drug deliverers
- Thermoplastic elastomers
- Compatibilizers for polymer blends
- Polymeric emulsifiers
- Impact resistant plastics

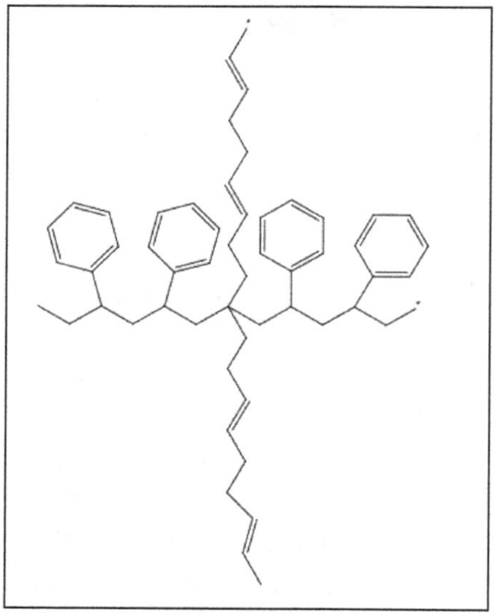

High Impact Polystyrene (HIPS) consists of the polystyrene backbone with polybutadiene chains branching from it in each direction.

High Impact Polystyrene

CD case made from general purpose polystyrene (GPPS) and high impact polystyrene in the black portion (HIPS).

High impact polystyrene (HIPS) was discovered by Charles F. Fryling in 1961. HIPS is a low cost, plastic material that is easy to fabricate and often used for low strength structural applications when impact resistance, machinability, and low cost are required. Its major applications include machined prototypes, low-strength structural components, housings, and covers. In order to produce the graft polymer, polybutadiene (rubber) or any similar elastomeric polymer is dissolved in styrene and polymerized. This reaction allows for two simultaneous polymerizations, that of styrene to polystyrene and that of the graft polymerization of styrene-rubber. During commercial use, it can be prepared by graft copolymerization with additional polymer to give the product specific characteristics. The advantages of HIPS includes:

- FDA compliant.
- Good impact resistance.
- Excellent machinability.
- Good dimensional stability.
- Easy to paint and glue.
- Low cost.
- Excellent aesthetic qualities.

New properties as a Result of Grafting

By grafting polymers onto polymer backbones, the final grafted copolymers gain new properties from their parent polymers. Specifically, cellulose graft copolymers have various different applications that are dependent on the structure of the polymer grafted onto the cellulose. Some of the new properties that cellulose gains from different monomers grafted onto it include:

- Absorption of water
- Improved elasticity
- Hydrophilic/Hydrophobic character
- Ion-exchange
- Dye adsorption capabilities
- Heat Resistance
- Thermosensitivity
- pH sensitivity
- Antibacterial effect

These properties give new application to the ungrafted cellulose polymers that include:

- Medical body fluid absorbent materials.
- Enhanced moisture absorbing ability in fabrics.
- Permselective membranes.
- Stronger nucleating properties than ungrafted cellulose, and adsorption of hazardous contaminants like heavy metal ions or dyes from aqueous solutions by temperature swing adsorption.
- Sensors and optical materials.
- Reducing agents for various carbonyl compounds.

Comb Polymer

Comb polymer, which consists of a hydrophobic poly(methyl methacrylate) (PMMA) backbone with hydrophilic hydroxy-poly(ethylene oxide) (HPOEM) side chains, is a tool that has many possible applications for the study of liver cell adhesion and signaling. This polymer has the unique properties of being cell resistant and chemically versatile such that various cell ligands can be coupled to its side chains. These properties allow adhesion through specific cell receptors to be studied without the effect of background adhesion to adsorbed proteins. By taking advantage of the ability to target specific receptors the comb polymer could be used as a powerful sorting tool. Sorting could be accomplished by finding cell type specific adhesion ligands. Several possible such ligands were screened. A ligand containing the tripeptide sequence RGD was found to elicit a strong cell adhesion response. However, this ligand is adherent to many cell types of the liver and would not be suitable for sorting purposes. Other cell type specific ligands tested showed little to no affinity for liver cell adhesion. Additionally, the comb was utilized to study $\alpha_5\beta_1$ integrin-specific hepatocyte adhesion and the effect of Epidermal Growth Factor on adhesion. $\alpha_5\beta_1$ integrin adhesion was mediated using a novel branched peptide, SynKRGD. This peptide consists of a linear peptide sequence containing RGDSP and the synergy site sequence PHSRN connected by the sequence GGKGGG. By utilizing the amine side group of Lysine a GGC branch was added. The terminal cysteine was used to conjugate SynKRGD to comb polymer surfaces using N-(p-Maleimidophenyl) isocyanate (PMPI) chemistry. EGF has a great potential to benefit the field of tissue engineering due to its influence on cell(cont.) proliferation, migration, and differentiation. EGF is also known to have a de-adhesive effect in some cell types. Hepatocytes were studied on comb surfaces of variable SynKRGD densities with and without the presence of EGF in the media. Distinct morphological differences were observed for hepatocytes on substrates of varying adhesivity with and without the presence of EGF. EGF was found to have a de-adhesive effect on $\alpha_5\beta_1$ integrin adhesion in hepatocytes. This effect became more pronounced as substrate adhesiveness increased.

Brush Polymer

A brush polymer is the name given to a surface coating consisting of polymers tethered to a surface. The brush may be either in a solvated state, where the tethered polymer layer consists of polymer and solvent, or in a melt state, where the tethered chains completely fill up the space available. These polymer layers can be tethered to flat substrates such as silicon wafers, or highly curved substrates such as nanoparticles. Also, polymers can be tethered in high density to another single polymer chain, although this arrangement is normally named a bottle brush.Additionally, there is a separate class of polyelectrolyte brushes, when the polymer chains themselves carry an electrostatic charge.

The brushes are often characterized by the high density of grafted chains. The limited space then leads to a strong extension of the chains. Brushes can be used to stabilize colloids, reduce friction between surfaces, and to provide lubrication in artificial joints.

Polymer brushes have been modeled with Monte Carlo methods, Brownian dynamics simulations, and molecular theories.

Structure

Polymer molecules within a brush are stretched away from the attachment surface as a result of the fact that they repel each other (steric repulsion or osmotic pressure). More precisely, they are more elongated near the attachment point and unstretched at the free end, as depicted on the drawing.

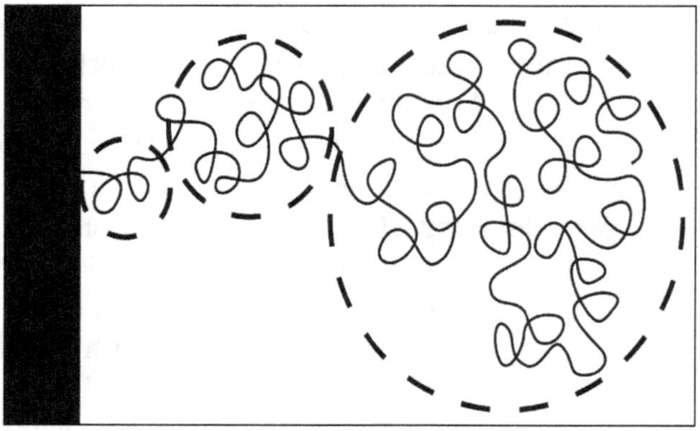

Polymer molecule within a brush. The drawing shows the chain elongation decreasing from the attachment point and vanishing at free end.

The "blobs", schematized as circles, represent the (local) length scale at which the statistics of the chain change from a 3D random walk (at smaller length scales) to a 2D in-plane random walk and a 1D normal directed walk (at larger length scales).

More precisely, within the approximation derived by Milner, Witten, Cates, the average density of all monomers in a given chain is always the same up to a prefactor:

$$\phi(z,\rho) = \frac{\partial n}{\partial z}$$

$$n(z,\rho) = \frac{2N}{\pi} \arcsin\left(\frac{z}{\rho}\right)$$

where ρ is the altitude of the end monomer and N the number of monomers per chain.

The averaged density profile $\epsilon(\rho)$ of the end monomers of all attached chains, convoluted with the above density profile for one chain, determines the density profile of the brush as a whole:

$$\phi(z) = \int_{\rho}^{\infty} \frac{\partial n(z,\rho)}{\partial z} \epsilon(\rho) \mathrm{d}\rho.$$

A dry brush has a uniform monomer density up to some altitude H. One can show that the corresponding end monomer density profile is given by:

$$\epsilon_{\text{dry}}(\rho,H) = \frac{\rho / H}{Na\sqrt{1-\rho^2/H^2}}$$

where a is the monomer size.

The above monomer density profile $n(z,\rho)$ for one single chain minimizes the total elastic energy of the brush,

$$U = \int_0^{\infty} \epsilon(\rho)\mathrm{d}\rho \int_0^{N} \mathrm{d}n \frac{kT}{2Na^2}\left(\frac{\partial z(n,\rho)}{\partial n}\right)^2.$$

regardless of the end monomer density profile $\epsilon(\rho)$.

From a Dry Brush to any Brush

As a consequence, the structure of any brush can be derived from the brush density profile $\phi(z)$. Indeed, the free end distribution is simply a convolution of the density profile with the free end distribution of a dry brush:

$$\epsilon(\rho) = \int_{\rho}^{\infty} -\frac{\mathrm{d}\phi(H)}{\mathrm{d}H} \epsilon_{\mathrm{dry}}(\rho,H)..$$

Correspondingly, the brush elastic free energy is given by:

$$\frac{F_{\mathrm{el}}}{kT} = \frac{\pi^2}{24N^2a^5} \int_0^{\infty} \left\{-z^3 \frac{\mathrm{d}\phi(z)}{\mathrm{d}z}\right\} \mathrm{d}z.$$

This method has been used to derive wetting properties of polymer melts on polymer brushes of the same species and to understand fine interpenetration asymmetries between copolymer lamellae that may yield very unusual non-centrosymmetric lamellar structures.

Dendronized Polymer

Dendronized polymers are formally a sub-group of comb polymers, where the linear chains of the grafts have been replaced by dendrons as depicted in Figure. The properties of these kinds of polymers are believed to depend on a number of factors, such as type of dendron, attachment density along the backbone, end-group functionality, and degree of polymerization. Because of the large number of properties that can be tailored independently in this type of architecture, several novel applications have been proposed, for example building blocks for nanometer sized construction, efficient light emitting materials, isolating conducting polymers, complexation agents for DNA, degradable drug carriers, and as support for catalysts. If spatially demanding dendrons are attached to a polymeric backbone, the polymer can be forced from a random-coil conformation to a stretched cylindrical.44,46 The multifunctional surface of these dendronized cylinders can be addressed and because of the potentially high bending modulus of these materials, dendronized polymers can be envisioned as molecular reinforcements in, for example, coating applications.

Synthesis of Dendronized Polymers

The first synthetic route to dendronized polymers was described in 1987 in a patent by Tomalia et al. and since then considerable synthetic effort has been devoted to developing new and efficient routs for these novel materials. Figure displays the two main routes to attaining dendronized polymers.

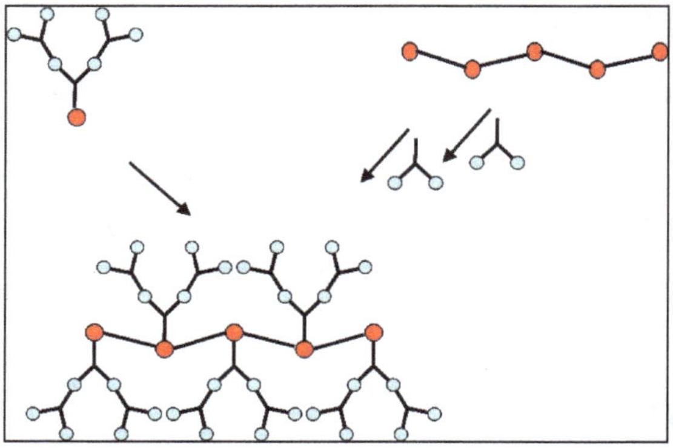

Figure: The macromonomer route (left), and the graft-onto route (right) to dendronized polymers.

In the macromonomer route, dendrons are linked to monomers that contain a polymerizable group either by a convergent or divergent methodology. The macromonomers are then polymerized, forming a dendronized polymer. methods have been employed in the macromonomer route for example radical polymerization, reversible addition-fragmentation chain transfer (RAFT) polymerization, ring-opening metathesis polymerization (ROMP), ring-opening polymerization (ROP), Suzuki polycondensation, insertion polymerization, Stille coupling, Heck coupling, polycondensation, and atom transfer radical polymerisation (ATRP).

The drawback of the macromonomer route is that larger dendrons may shield the polymerizable group and thereby inhibit the polymerization, resulting in either low molecular weight polymers or no reaction at all. Studies have addressed this problem, and propose that by increasing the distance between dendron and polymerizable group, the availability of the reactive group will increase. However, recent studies indicate that the problem of inhibition may be related only to the concentration of the monomer, if the concentration of monomer is kept high enough, high degrees of polymerization can be reached.

In the 'graft-onto' route preformed dendrons are attached to a polymeric backbone. Both convergent and divergent approaches have been employed. The 'graft-onto' route may suffer the drawback of incomplete coupling steps due to steric crowding of the dendrons. To circumvent this problem, large excesses of coupling agent and long reaction times must be employed. The large excesses may lead to purification problems of the product. Divergent approaches to dendronized polymers utilizing either benzylidene-2,2- bis(methoxy) propanoic anhydride or acetonide-2,2-bis(methoxy) propanoic anhydride allow for purification of large excesses of coupling agent by simple extractions. The benefit of combining the 'graft-onto' route with a controlled polymerization technique is that narrowly distributed dendronized polymers emanating from the same backbone can be produced with ease.

Structure and Applications

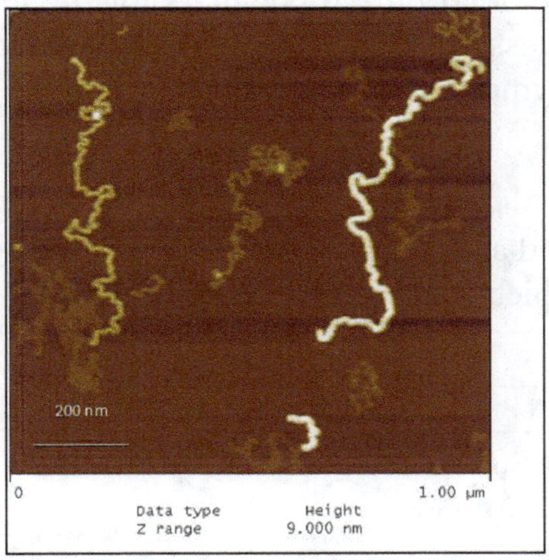

Figure: Atomic force microscopy height image of co-prepared dendronized polymers of generation one through four (PG1-PG4) reflecting the different thicknesses and apparent persistence lengths for each generation.

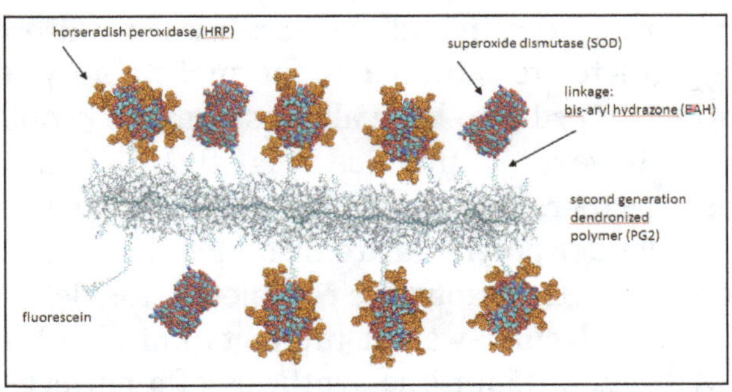

Figure: Schematic representation of a molecular hybrid structure (conjugate) between a dendronized polymer and the two different enzymes HRP (horseradish peroxidase) and SOD (Cu,Zn-superoxide dismutase). PDB (SOD): 1SXA; PDB (HRP): 1ATJ; HRP sugar modification from Gray & Montgomery, *Carbohydrate*.

Dendronized polymers can contain several thousands of dendrons in one macromolecule and have a stretched out, anisotropic structure. In this regard they differ from the more or less spherically shaped dendrimers, where a few dendrons are attached to a small, dot-like core resulting in an isotropic structure. Depending on dendron generation, the polymers differ in thickness as the atomic force microscopy image shows Figure. Neutral and charged dendronized polymers are highly soluble in organic solvents and in water, respectively. This is due to their low tendency to entangle. Dendronized polymers have been synthesized with, e.g., polymethylmethacrylate, polystyrene, polyacetylene, polyphenylene, polythiophene, polyfluorene, poly(phenylene vinylene), poly(phenylene acetylene), polysiloxane, polyoxanorbornene, poly(ethylene imine) (PEI) backbones. Molar masses up to 200 Mio g/mol have been obtained. Dendronized polymers have been investigated for/as bulk structure control, responsivity to external stimuli, single molecule chemistry, templates for nanoparticle formation, catalysis, electro-optical devices, and bio-related applications. Particularly attractive is the use of water-soluble dendronized polymers for the immobilization of enzymes on solid surfaces (inside glass tubes or microfluidic devices) and for the preparation of dendronized polymer-enzyme conjugates.

Ladder Polymers

A ladder polymer is defined as a "uninterrupted series of rings connected by sterically restrictive links around which rotation cannot occur without the breaking of a bond" .

"Classical" ladder Polymer

"Imperfect" ladder or step ladder polymer

Conjugated ladder polymers possess a planarized π-electron system which ensures optimum electron delocalization. Theory predicts that such materials will exhibit outstanding electronic properties including siginificant third-order susceptibilities. Ladder polymers are predicted to show a high thermal stability and a high resistance to chemical degradation. In order to have all these unique properties perfect ladder polymers are required. However, in practice, it is often difficult to bring nearly perfect ladder structures for experimental realization. There are many reasons for these difficulties. First, many ladder polymers are insoluble in common solvents and precipitate from solutions before reactions are complete. This deficiency not only leads to the formation of low molecular-weight products but also leads to incompletely cyclized structures. Without a doubt, the synthesis of a perfect ladder polymer is a real synthetic challenge. In the case of ladder polymers, two independent strands have to be generated without defects or crosslinking. In the early 1970s poly(benzimidazobenzophenanthroline) (BBL 17) was synthesized and it was discovered that this ladder polymer is readily processible into durable films using casting or spraying techniques.

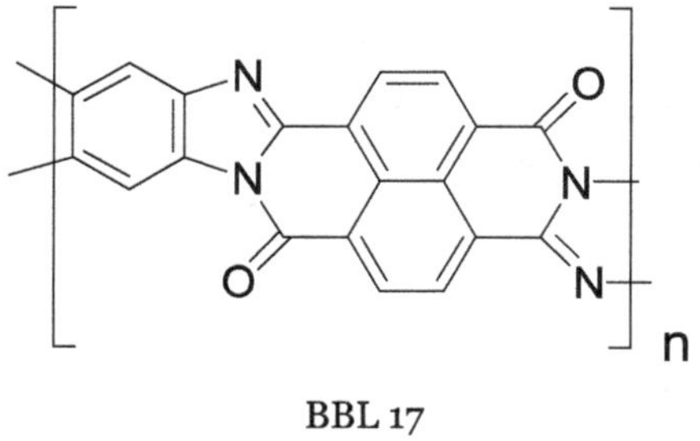

BBL 17

Several other ladder polymers were than synthesized and a few ladder polymers gained some industrial importance as materials for high-temperature applications such as heat-resistant coatings e.g. in aerospace vehicles.

Synthesis of Ladder Polymers

Attempts to generate ladder polymers have a long history, but ladder polymers prepared at the initial stage showed a poor solubility, due to their rigid structure. Many synthetic efforts have been made to bring soluble ladder polymer to an experimental realization. A new generation of substituted, soluble ladder polymer have been made using the following improved approaches:

- Novel polycondensation methods e.g. repetative Diels-alder reaction.
- Polymer-analogous cyclization of single-stranded linear prepolymers.

Most of the known synthesis of ladder polymers fall into the these two methods. But it is essential to prevent side-reactions and to take care of an almost quantitative conversion of the starting materials. It is beyond the scope of this chapter to cover all the history of ladder polymers. We will deal with the conjugated poly(para-phenylene)-type (LPPP) ladder polymers. The polymer-analogous cyclization of suitable single-stranded precursors have been successfully applied to synthesize such polyphenylene-based conjugated ladder polymers.

Two-dimensional Polymers

Two-dimensional polymers are sheet-like, covalently bonded molecular tilings that extend in exactly two dimensions. Unlike 1D polymers, over which chemists can often exert high levels of control during their synthesis, the design and production of 2D polymers presents far greater challenges. Graphene is perhaps the archetypal 2D material, and although it has enjoyed a great deal of attention, the idea of precisely tailoring its size, shape and functionality at the molecular level seems somehow incompatible with something that can be made in a kitchen blender. Perusing the recent literature, you could be forgiven for thinking that graphene was the only 2D material in existence, but other natural and synthetic layered substances have also been exfoliated into 2D sheets.

The unique optical and electronic properties that arise from the 2D structure of graphene have attracted the attention of many researchers, including those interested in modulating its behaviour through chemical functionalization — although there can be some uncertainty regarding the exact nature and placement of functional groups. Bottom-up approaches to functional graphene fragments5 (or so-called nanographenes) look to overcome the problems associated with top-down methods, but are currently limited in terms of the size of sheets that can be formed, and the functionality that can be introduced. Two-dimensional synthetic polymers arguably have the potential to overcome these limitations, particularly if large crystals can be grown.

References

- Polymer-Architecture, chemical-engineering, technical-references: idconline.com, Retrieved 22, May 2020
- Polymer-sciences-journals-articles-ppts-list: omicsonline.org, Retrieved 7, July 2020
- Xie, Jiangbing; Hsieh, You-Lo (25 July 2003). "Thermosensitive poly(n-isopropylacrylamide) hydrogels bonded on cellulose supports". Journal of Applied Polymer Science. 89 (4): 999–1006. doi:10.1002/app.12206
- Star-copolymers, polymers-with-Specific-Architecture: cmu.edu, Retrieved 2, April 2020
- Feng, Chun; Li, Yongjun; Yang, Dong; Hu, Jianhua; Zhang, Xiaohuan; Huang, Xiaoyu (2011). "Well-defined graft copolymers: from controlled synthesis to multipurpose applications". Chemical Society Reviews. 40 (3): 1282–95. doi:10.1039/b921358a. PMID 21107479
- editors, Susheel Kalia, M.W. Sabaa, (2013). Polysaccharide based graft copolymers (1., 2013 ed.). Heidelberg: Springer. ISBN 9783642365652
- Campbell, Neil A.; Brad Williamson; Robin J. Heyden (2006). Biology: Exploring Life. Boston, Massachusetts: Pearson Prentice Hall. ISBN 0-13-250882-6
- Volker Abetz ... [et (2005). Encyclopedia of polymer science and technology (Wird aktualisiert. ed.). [Hoboken, N.J.]: Wiley-Interscience. ISBN 9780471440260
- Future-scope-of-polymer-science: polymerchemistry.chemistryconferences.org, Retrieved 7, May 2020
- Pearce, Eli M. (May 1987). "New commercial polymers 2, by Hans-George Elias and Friedrich Vohwinkel, Gordon and Breach, New York, 1986, 508 pp. Price: $90.00". Journal of Polymer Science Part C: Polymer Letters. 25 (5): 233–234. doi:10.1002/pol.1987.140250509

6

Significant Theories in Polymer Chemistry

Flory-Stockmayer theory predicts the gel point of the system containing three types of monomer units; linear units with two A-groups, linear units with two B groups and branched A units. Flory-Huggins solution theory is the most prominent method for understanding phase behavior of polymer mixtures and thermodynamics. This chapter discusses in detail the theories and methodologies related to polymer chemistry.

Flory-Stockmayer Theory

Flory-Stockmayer Theory is a theory governing the cross-linking and gelation of step-growth polymers:.. The Flory-Stockmayer theory represents an advancement from the Carothers equation, allowing for the identification of the gel point for polymer synthesis not at stoichiometric balance. The theory was initially conceptualized by Paul Flory in 1941 and then was further developed by Walter Stockmayer in 1944 to include cross-linking with an arbitrary initial size distribution.

History

Gelation occurs when a polymer forms large interconnected polymer molecules through cross-linking. In other words, polymer chains are cross-linked with other polymer chains to form an infinitely large molecule, interspersed with smaller complex molecules, shifting the polymer from a liquid to a network solid or gel phase. The Carothers equation is an effective method for calculating the degree of polymerization for stoichiometrically balanced reactions. However, the Carothers equation is limited to branched systems, describing the degree of polymerization only at the onset of cross-linking. The Flory-Stockmayer Theory allows for the prediction of when gelation occurs using percent conversion of initial monomer and is not confined to cases of stoichiometric balance. Additionally, the Flory-Stockmayer Theory can be used to predict whether gelation is possible through analyzing the limiting reagent of the step-growth polymerization.

Flory's Assumptions

In creating the Flory-Stockmayer Theory, Flory made three assumptions that affect the accuracy of this model. These assumptions were,

1. All functional groups on a branch unit are equally reactive.
2. All reactions occur between A and B.
3. There are no intramolecular reactions.

As a result of these assumptions, a conversion slightly higher than that predicted by the Flory-Stockmayer Theory is commonly needed to actually create a polymer gel. Since steric hindrance effects prevent each functional group from being equally reactive and intramolecular reactions do occur, the gel forms at slightly higher conversion.

General Case

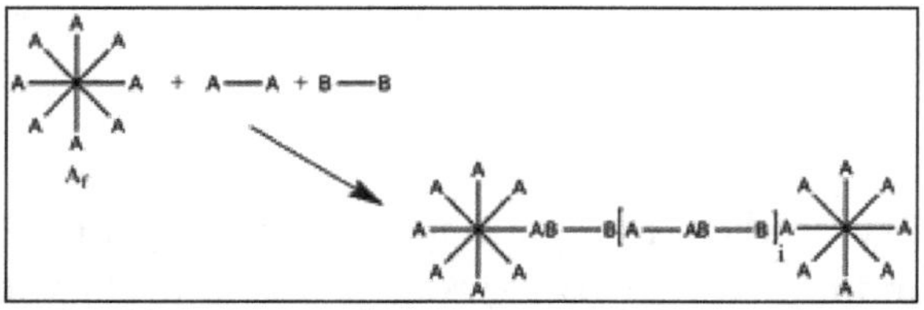

A general image of a multifunctional branch unit, A_f, reacting with bifunctional monomers with A and B functional groups to form a step-growth polymer.

The Flory-Stockmayer Theory predicts the gel point for the system consisting of three types of monomer units:

linear units with two A-groups (concentration c_1),

linear units with two B groups (concentration c_2),

branched A units (concentration c_3).

The following definitions are used to formally define the system:

f is the number of reactive functional groups on the branch unit (i.e. the functionality of that branch unit)

p_A is the probability that A has reacted (conversion of A groups)

p_B is the probability that B has reacted (conversion of B groups)

$\rho = \frac{fc_3}{2c_1 + fc_3}$ is the ratio of number of A groups in the branch unit to the total number of A groups

$r = \frac{2c_1 + fc_3}{2c_2} = \frac{p_B}{p_A}$ is the ratio between total number of A and B groups. So that $p_B = rp_A$.

The theory states that the gelation occurs only if $\alpha > \alpha_c$, where

$$\alpha_c = \frac{1}{f-1}$$

is the critical value for cross-linking and α is presented as a function of p_A,

$$\alpha(p_A) = \frac{rp_A^2\rho}{1 - rp_A^2(1-\rho)}$$

or, alternatively, as a function of p_B,

$$\alpha(p_B) = \frac{p_B^2 \rho}{r - p_B^2(1-\rho)}.$$

One may now substitute expressions for r, ρ into definition of α and obtain the critical values of $p_A, (p_B)$ that admit gelation. Thus gelation occurs if:

$$p_A > \sqrt{\frac{\alpha_c}{r(\alpha_c + \rho - \alpha_c \rho)}}.$$

alternatively, the same condition for p_B reads,

$$p_B > \sqrt{\frac{r\alpha_c}{\alpha_c + \rho - \alpha_c \rho}}$$

The both inequalities are equivalent and one may use the one that is more convenient. For instance, depending on which conversion p_A or p_B is resolved analytically.

Trifunctional A Monomer with Difunctional B Monomer

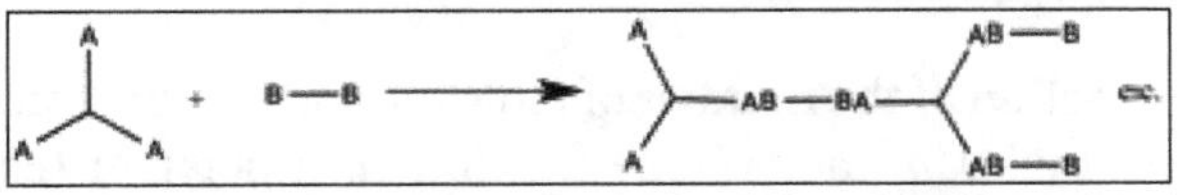

A trifunctional branch unit with A functional group reacting with a bifunctional branch unit with a B functional group, forming a continuous step-growth polymer molecule.

$$\alpha_c = \frac{1}{f-1} = \frac{1}{3-1} = \frac{1}{2}$$

Since all the A functional groups are from the trifunctional monomer, ρ = 1 and

$$\alpha = \frac{\frac{p_B^2 \rho}{r}}{1 - \frac{p_B^2}{r(1-\rho)}} = \frac{p_B^2}{r}$$

Therefore, gelation occurs when

$$\frac{p_B^2}{r} > \alpha_c$$

or when,

$$p_B > \sqrt{\frac{r}{2}}$$

Similarly, gelation occurs when

$$p_A > \sqrt{\frac{1}{2r}}$$

Flory–Huggins Solution Theory

Flory–Huggins solution theory is a mathematical model of the thermodynamics of polymer solutions which takes account of the great dissimilarity in molecular sizes in adapting the usual expression for the entropy of mixing. The result is an equation for the Gibbs free energy change ΔG_m for mixing a polymer with a solvent. Although it makes simplifying assumptions, it generates useful results for interpreting experiments.

The thermodynamic equation for the Gibbs energy change accompanying mixing at constant temperature and (external) pressure is,

$$\Delta G_m = \Delta H_m - T\Delta S_m$$

A change, denoted by Δ, is the value of a variable for a solution or mixture minus the values for the pure components considered separately. The objective is to find explicit formulas for ΔH_m and ΔS_m, the enthalpy and entropy increments associated with the mixing process.

The result obtained by Flory and Huggins is,

$$\Delta G_m = RT[n_1 \ln \phi_1 + n_2 \ln \phi_2 + n_1 \phi_2 \chi_{12}]$$

The right-hand side is a function of the number of moles n_1 and volume fraction ϕ_1 of solvent (component 1), the number of moles n_2 and volume fraction ϕ_2 of polymer (component 2), with the introduction of a parameter χ to take account of the energy of interdispersing polymer and solvent molecules. R is the gas constant and T is the absolute temperature. The volume fraction is analogous to the mole fraction, but is weighted to take account of the relative sizes of the molecules. For a small solute, the mole fractions would appear instead, and this modification is the innovation due to Flory and Huggins. In the most general case the mixing parameter, χ, is a free energy parameter, thus including an entropic component.

Derivation

We first calculate the *entropy* of mixing, the increase in the uncertainty about the locations of the molecules when they are interspersed. In the pure condensed phases — solvent and polymer — everywhere we look we find a molecule. Of course, any notion of "finding" a molecule in a given location is a thought experiment since we can't actually examine spatial locations the size of molecules. The expression for the entropy of mixing of small molecules in terms of mole fractions is no longer reasonable when the solute is a macromolecular chain. We take account of this dissymmetry in molecular sizes by assuming that individual polymer segments and individual solvent molecules occupy sites on a lattice. Each site is occupied by exactly one molecule of the solvent or by one monomer of the polymer chain, so the total number of sites is,

$$N = N_1 + xN_2$$

N_1 is the number of solvent molecules and N_2 is the number of polymer molecules, each of which has x segments.

From statistical mechanics we can calculate the entropy change, the increase in spatial uncertainty, as a result of mixing solute and solvent.

$$\Delta S_m = -k[N_1 \ln(N_1 / N) + N_2 \ln(xN_2 / N)]$$

where k is Boltzmann's constant. Define the lattice *volume fractions* ϕ_1 and ϕ_2,

$$\phi_1 = N_1 / N$$

$$\phi_2 = xN_2 / N$$

These are also the probabilities that a given lattice site, chosen at random, is occupied by a solvent molecule or a polymer segment, respectively. Thus,

$$\Delta S_m = -k[N_1 \ln \phi_1 + N_2 \ln \phi_2]$$

For a small solute whose molecules occupy just one lattice site, x equals one, the volume fractions reduce to molecular or mole fractions, and we recover the usual equation from ideal mixing theory.

In addition to the entropic effect, we can expect an *enthalpy* change. There are three molecular interactions to consider: solvent-solvent w_{11}, monomer-monomer w_{22} (not the covalent bonding, but between different chain sections), and monomer-solvent w_{12}. Each of the last occurs at the expense of the average of the other two, so the energy increment per monomer-solvent contact is,

$$\Delta w = w_{12} - \frac{1}{2}(w_{22} + w_{11})$$

The total number of such contacts is

$$xN_2 z\phi_1 = N_1\phi_2 z$$

where z is the coordination number, the number of nearest neighbors for a lattice site, each one occupied either by one chain segment or a solvent molecule. That is, xN_2 is the total number of polymer segments (monomers) in the solution, so $xN_2 z$ is the number of nearest-neighbor sites to *all* the polymer segments. Multiplying by the probability ϕ_1 that any such site is occupied by a solvent molecule, we obtain the total number of polymer-solvent molecular interactions. An approximation following mean field theory is made by following this procedure, thereby reducing the complex problem of many interactions to a simpler problem of one interaction.

The enthalpy change is equal to the energy change per polymer monomer-solvent interaction multiplied by the number of such interactions,

$$\Delta H_m = N_1\phi_2 z \Delta w$$

The polymer-solvent interaction parameter *chi* is defined as,

$$\chi_{12} = z\Delta w / kT$$

It depends on the nature of both the solvent and the solute, and is the only *material-specific* parameter in the model. The enthalpy change becomes,

$$\Delta H_m = kTN_1\phi_2\chi_{12}$$

Assembling terms, the total free energy change is,

$$\Delta G_m = RT[n_1 \ln \phi_1 + n_2 \ln \phi_2 + n_1\phi_2\chi_{12}]$$

where we have converted the expression from molecules N_1 and N_2 to moles n_1 and n_2 by transferring Avogadro's number N_A to the gas constant $R = kN_A$.

The value of the interaction parameter can be estimated from the Hildebrand solubility parameters δ_a and δ_b,

$$\chi_{12} = V_{seg}(\delta_a - \delta_b)^2 / RT$$

where V_{seg} is the actual volume of a polymer segment.

This treatment does not attempt to calculate the conformational entropy of folding for polymer chains. The conformations of even amorphous polymers will change when they go into solution, and most thermoplastic polymers also have lamellar crystalline regions which do not persist in solution as the chains separate. These events are accompanied by additional entropy and energy changes.

It should be noted that in the most general case the interaction Δw and the ensuing mixing parameter, χ, is a free energy parameter, thus including an entropic component. This means that aside to the regular mixing entropy there is another entropic contribution from the interaction between solvent and monomer. This contribution is sometimes very important in order to make quantitative predictions of thermodynamic properties.

More advanced solution theories exist, such as the Flory-Krigbaum theory.

Oligomer-gel Solutions Theory

Recently a team of scientists from the University of Durham (UK) introduced a new form of the Flory-Huggins free energy for a solution where there is an oligomer with a cross-linked polymer like a reticulated gel. They considered the mixing entropy for the oligomer and the elastic entropy for the reticulated gel keeping the mean field interaction through χ. They found that increasing the elastic modulus of the cross-linked polymer it is possible to increase the critical value of the mean field over which there is phase separation. This theory introduces a new way of controlling the phase separation in such a system. Furthermore they used this theory for studying the wetting and the migration of the oligomer specie to a surface.

Hoffman Nucleation Theory

Hoffman nucleation theory is a theory developed by John D. Hoffman and coworkers in the 1970s and 80s that attempts to describe the crystallization of a polymer in terms of the kinetics and thermodynamics of polymer surface nucleation. The theory introduces a model where a surface of completely crystalline polymer is created and introduces surface energy parameters to describe the process. Hoffman nucleation theory is more of a starting point for polymer crystallization theory and is better known for its fundamental roles in the Hoffman-Weeks Lamellar Thickening and Lauritzen-Hoffman Growth Theory.

Polymer Morphology

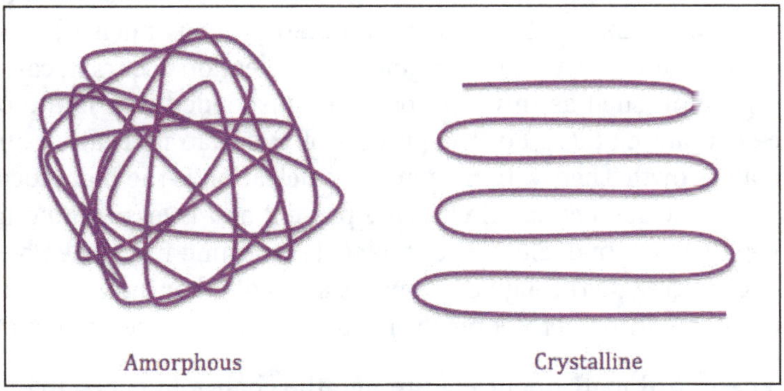

Amorphous regions lack the energy needed to order into folded regions such as those seen in the Crystalline state.

Polymers contain different morphologies on the molecular level which give rise to their macro properties. Long range disorder in the polymer chain is representative of amorphous solids, and the chain segments are considered amorphous. Long range polymer order is similar to crystalline material, and chain segments are considered crystalline.

The thermal characteristics of polymers are fundamentally different from those of most solid materials. Solid materials typically have one melting point, the T_m, above which the material loses internal molecular ordering and becomes a liquid. Polymers have both a melting temperature T_m and a glass transition temperature T_g. Above the T_m, the polymer chains lose their molecular ordering and exhibit reptation, or mobility. Below the T_m, but still above the T_g, the polymer chains lose some of their long-range mobility and can form either crystalline or amorphous regions. In this temperature range, as the temperature decreases, amorphous regions can transition into crystalline regions, causing the bulk material to become more crystalline over all. Below the T_g, molecular motion is stopped and the polymer chains are essentially frozen in place. In this temperature range, amorphous regions can no longer transition into crystalline regions, and the polymer as a whole has reached its maximum crystallinity.

	Temperature	Polymer Macroproperties	Polymer Morphology
	$>T_m$	Liquid	No long-range order, polymer chains are fully amorphous and can reptate.
T_m	$T_m > T > T_g$	Rubber	Polymer chains contain amorphous and crystalline regions and lower overall mobility.
T_g	$<T_g$	Glassy/Brittle	Polymer chains are stuck in place and no more crystallinity can be achieved.

Hoffman nucleation theory addresses the amorphous to crystalline polymer transition, and this transition can only occur in the temperature range between the T_m and T_g. The transition from an amorphous to a crystalline single polymer chain is related to the random thermal energy required to align and fold sections of the chain to form ordered regions titled lamellae, which are a subset of even bigger structures called spherulites. The Crystallization of polymers can be brought about by several different methods, and is a complex topic in itself.

Nucleation

Nucleation is the formation and growth of a new phase with or without the presence of external surface. The presence of this surface results in heterogeneous nucleation whereas in its absence homogeneous nucleation occurs. Heterogeneous nucleation occurs in cases where there are pre-existing nuclei present, such as tiny dust particles suspended in a liquid or gas or reacting with a glass surface containing SiO_2. For the process of Hoffman nucleation and its progression to Lauritzen-Hoffman Growth Theory, homogeneous nucleation is the main focus. Homogeneous nucleation occurs where no such contaminants are present and is less commonly seen. Homogeneous nucleation begins with small clusters of molecules forming from one phase to the next. As the clusters grow, they aggregate through the condensation of other molecules. The size continues to increase and ultimately form macroscopic droplets (or bubbles depending on the system).

Nucleation is often described mathematically through the change in Gibbs free energy of n moles of vapor at vapor pressure P that condenses into a drop. Also the nucleation barrier, in polymer crystallization, consists of both enthalpic and entropic components that must be over come. This barrier consists of selection processes taking place in different length and time scales which relates to the multiple regimes later on. This barrier is the free energy required to overcome in order to form nuclei. It is the formation of the nuclei from the bulk to a surface that is the interfacial free energy. The interfacial free energy is always a positive term and acts to destabilize the nucleus allowing the continuation of the growing polymer chain. The nucleation continues as a favorable reaction.

Thermodynamics of Polymer Crystallization

The Lauritzen–Hoffman plot (right) models the three different regimes when (logG) + $U^*/k(T-T_0)$ is plotted against $(T\Delta T)^{-1}$. It can be used to describe the rate at which secondary nucleation competes with lateral addition at the growth front among the different temperatures. This theory can be used to help understand the preferences of nucleation and growth based on the polymer's properties including its standard melting temperature.

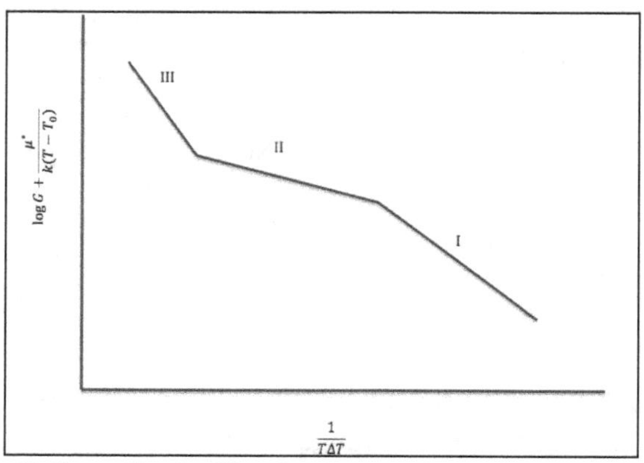

Lauritzen–Hoffman Plot detailing the Three Regimes of Secondary Nucleation.

Lamellar Thickening (Hoffman–Weeks plot)

For many polymers, the change between the initial lamellar thickness at T_c is roughly the same as at T_m and can thus be modeled by the Gibbs–Thomson equation fairly well. However, since it implies that the lamellar thickness over the given supercooling range (T_m-T_c) is unchanged, and many homogeneous nucleation of polymers implies a change of thickness at the growth front, Hoffman and Weeks pursued a more accurate representation. In this regard, the Hoffman-Weeks plot was created and can be modeled through the equation,

$$T_m = \frac{T_c}{\beta} + (1 - \frac{1}{\beta})T_m^{\circ}$$

where β is representative of a thickening factor given by $L = L_o\ \beta$ and T_c and T_m are the crystallization and melting temperatures, respectively.

Applying this experimentally for a constant β allows for the determination of the equilibrium melting temperature, T_m° at the intersection of T_c and T_m.

Kinetics of Polymer Crystallization

The crystallization process of polymers does not always obey simple chemical rate equations. Polymers can crystallize through a variety of different regimes and unlike simple molecules, the polymer crystal lamellae have two very different surfaces. The two most prominent theories in polymer crystallization kinetics are the Avrami equation and Lauritzen-Hoffman Growth Theory.

Lauritzen-Hoffman Growth Theory

The Lauritzen-Hoffman growth theory breaks the kinetics of polymer crystallization into ultimately two rates. The model breaks down into the addition of monomers onto a growing surface. This initial step is generally associated with the nucleation of the polymer. From there, the kinetics become the rate which the polymer grows on the surface, or the lateral growth rate, in comparison with the growth rate onto the polymer extending the chain, the secondary nucleation rate. These two rates can result in three situations.

Three Regimes of Crystallization Kinetics

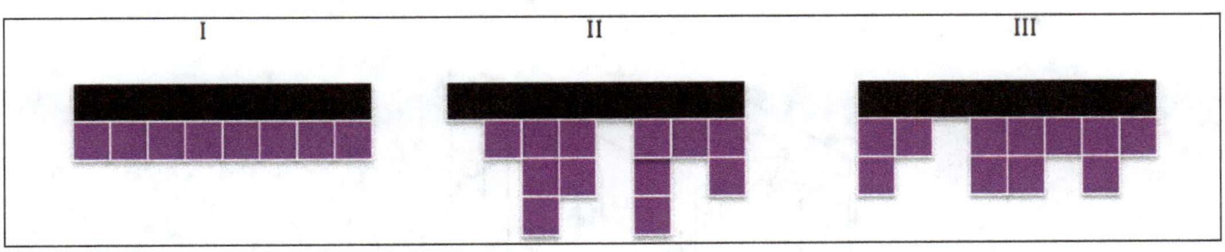

For Regime I, the growth rate on the front laterally, referred to as *g*, is the rate-determining step (RDS) and exceeds the secondary nucleation rate, *i*. In this instance of $g >> i$, monolayers are formed one at a time so that if the substrate has a length of L_p and thickness, *b*, the overall linear growth can be described through the equation,

$$G_I = biL_p$$

and the rate of nucleation in specific can further be described by,

$$G_{\mathrm{I,n}} = e^{-(K_g/T\Delta T)}$$

with K_g equal to,

$$K_g = \frac{4b\sigma_1\sigma_f T_m^0}{k\Delta h}$$

where:

- σ_l is the lateral/lamellae surface free energy per unit area
- σ_f is the fold surface free energy per unit area
- T_m^o is the equilibrium melting temperature
- k is equal to Boltzmann Constant
- Δh is equal to the change of enthalpy of fusion per repeat unit at the standard temperature (Also known as the Latent Heat of Fusion per repeat unit)

This shows that in Region I, lateral nucleation along the front successfully dominates at temperatures close to the melting temperature, however at more extreme temperatures other forces such as diffusion can impact nucleation rates.

In Regime II, the lateral growth rate is either comparable or smaller than the nucleation rate $g \leq i$, which causes secondary (or more) layers to form before the initial layer has been covered. This allows the linear growth rate to be modeled by,

$$G_{\mathrm{II}} = b\sqrt{ig}$$

Using the assumption that g and i are independent of time, the rate at which new layers are formed can be approximated and the rate of nucleation in regime II can be expressed as:

$$G_{\mathrm{II,n}} = e^{-(K_g'/T\Delta T)}$$

with K_g' equal to about 1/2 of the K_g from Regime I,

$$K_g' = \frac{2b\sigma_1\sigma_f T_m^0}{k\Delta h}$$

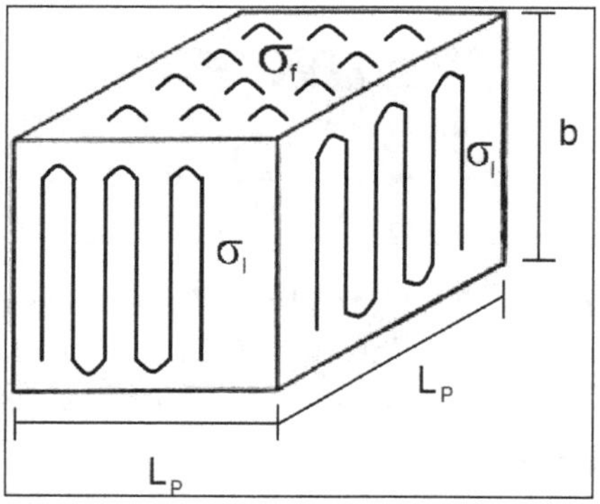

Diagram of a crystalline polymer lamellae.

Lastly, Regime III in the L-H model depicts the scenario where lateral growth is inconsequential to the overall rate, since the nucleation of multiple sites causes $i >> g$. This means that the growth rate can be modeled by the same equation as Regime I,

$$G_{\text{III}} = biL_{\text{p}} = G_{\text{III}}^{\circ} e^{-U^{*}/k(T-T_0)-(K_g/T\Delta T)}$$

where G_{III}° is the prefactor for Regime III and can be experimentally determined through applying the Lauritzen–Hoffman Plot.

Polyethylene Crystallization Kinetics

A reza's crystallization depends on the time it takes for layers of its chains to fold and orient themselves in the same direction. This time increases with a molecule's weight and branching. The table below shows that the growth rate is higher for Sclair 14B.1 than Sclair 2907 (20%), where 2907 is less highly branched than 14B.1. Here Gc is the crystal growth rate, or how quickly it orders itself depending on the layers, and t is the time it takes to order.

Polymer	Growth Temp (°C)	G_c (μm*min^{-1})	t (ms)
Sclair 2907 (20%)	119	3.5-6.8	4.4-8.6
Sclair 14B.1	119	~0.2	~150

Further Testing and Applications

Many additional tests have since been run to apply and compare Hoffman's principles to reality. Among the experiments done, some of the more notable secondary nucleation tests are briefly explained in the table below.

Secondary nucleation testing	Experimental results observed
Potassium chloride (KCl)	Secondary nuclei form at a rate proportional to the degree of supercooling (above certain levels of agitation) and achieve the same amount of nucleation regardless of the shape of the parent crystal. This is due to the substantially greater effect of secondary nucleation over primary nucleation of the original crystal. This was proven through both temperature and shape dependent nuclei stimulated growth experiments to confirm that in cases of secondary nucleation only the degree and temperature of supercooling change the nucleation rate, whereas the parent crystal only serves to act as a catalytic initiator of the process.
Isotactic poly (vinylcyclohexane) (PVCH)	PVCH crystals were experimentally shown to increase their spread and lateral growth at high temperatures indicating that although they were feasibly unable to reach Regime III temperatures, extrapolations and hypotheses from the experiment point to confirmation of the expected behavior in each of the three regimes. Experiments concluded that additional growth mechanisms such as crystal twinning and twin boundary interactions can alter the traditional LH theory, but further research is needed to model each individual influence.

Zinc oxide (ZnO)	Zinc oxide crystals were proven to undergo secondary nucleation under an odd mixture of conditions including the addition of a diamine as well as surface etching. Overall, testing has shown that the morphology of the secondary crystals can fluctuate greatly depending on the amount of diamine added, due to its ability to deplete the substrate and hinder the growth prematurely.

Polymer Field Theory

A polymer field theory is a statistical field theory describing the statistical behavior of a neutral or charged polymer system. It can be derived by transforming the partition function from its standard many-dimensional integral representation over the particle degrees of freedom in a functional integral representation over an auxiliary field function, using either the Hubbard-Stratonovich transformation or the delta-functional transformation. Computer simulations based on polymer field theories have been shown to deliver useful results, for example to calculate the structures and properties of polymer solutions (Baeurle 2007, Schmid 1998), polymer melts (Schmid 1998, Matsen 2002, Fredrickson 2002) and thermoplastics (Baeurle 2006).

Canonical Ensemble

Particle Representation of the Canonical Partition Function

The standard continuum model of flexible polymers, introduced by Edwards (Edwards 1965), treats a solution composed of n linear monodisperse homopolymers as a system of coarse-grained polymers, in which the statistical mechanics of the chains is described by the continuous Gaussian thread model (Baeurle 2007) and the solvent is taken into account implicitly. The Gaussian thread model can be considered as the continuum limit of the discrete Gaussian chain model, in which the polymers are described as continuous, linearly elastic filaments. The canonical partition function of such a system, kept at an inverse temperature $\beta = 1/k_B T$ and confined in a volume V, can be expressed as:

$$Z(n,V,\beta) = \frac{1}{n!(\lambda_T^3)^{nN}} \prod_{j=1}^{n} \int D\mathbf{r}_j \exp\left(-\beta\Phi_0[\mathbf{r}] - \beta\bar{\Phi}[\mathbf{r}]\right), \qquad (1)$$

where $\bar{\Phi}[\mathbf{r}]$ is the potential of mean force given by,

$$\bar{\Phi}[\mathbf{r}] = \frac{N^2}{2} \sum_{j=1}^{n} \sum_{k=1}^{n} \int_0^1 ds \int_0^1 ds' \bar{\Phi}\left(\left|\mathbf{r}_j(s) - \mathbf{r}_k(s')\right|\right) - \frac{1}{2} nN\bar{\Phi}(0), \qquad (2)$$

representing the solvent-mediated non-bonded interactions among the segments, while $\Phi_0[\mathbf{r}]$ represents the harmonic binding energy of the chains. The latter energy contribution can be formulated as:

$$\Phi_0[\mathbf{r}] = \frac{3k_B T}{2Nb^2} \sum_{l=1}^{n} \int_0^1 ds \left|\frac{d\mathbf{r}_l(s)}{ds}\right|^2,$$

where b is the statistical segment length and N the polymerization index.

Field-theoretic Transformation

To derive the basic field-theoretic representation of the canonical partition function, one introduces in the following the segment density operator of the polymer system,

$$\hat{\rho}(\mathbf{r}) = N\sum_{j=1}^{n}\int_{0}^{1} ds\delta\left(\mathbf{r}-\mathbf{r}_j(s)\right).$$

Using this definition, one can rewrite Eq. (2) as:

$$\bar{\Phi}[\mathbf{r}] = \frac{1}{2}\int d\mathbf{r}\int d\mathbf{r}'\hat{\rho}(\mathbf{r})\bar{\Phi}(|\mathbf{r}-\mathbf{r}'|)\hat{\rho}(\mathbf{r}') - \frac{1}{2}nN\bar{\Phi}(0). \qquad (3)$$

Next, one converts the model into a field theory by making use of the Hubbard-Stratonovich transformation or delta-functional transformation,

$$\int D\rho\, \delta[\rho-\hat{\rho}]F[\rho] = F[\hat{\rho}], \qquad (4)$$

where $F[\hat{\rho}]$ is a functional and $\delta[\rho \quad \hat{\rho}]$ is the delta functional given by:

$$\delta[\rho-\hat{\rho}] = \int Dw e^{i\int d\mathbf{r} w(\mathbf{r})[\rho(\mathbf{r})-\hat{\rho}(\mathbf{r})]}, \qquad (5)$$

with $w(\mathbf{r}) = \sum_{\mathbf{G}} w(\mathbf{G})\exp[i\mathbf{G}\mathbf{r}]$ representing the auxiliary field function. Here we note that, ex panding the field function in a Fourier series, implies that periodic boundary conditions are applied in all directions and that the $\mathbf{G}$-vectors designate the reciprocal lattice vectors of the supercell.

Basic field-theoretic Representation of Canonical Partition Function

Using the Eqs. (3), (4) and (5), we can recast the canonical partition function in Eq. (1) in field-theoretic representation, which leads to,

$$Z(n,V,\beta) = Z_0\int Dw\exp\left[-\frac{1}{2\beta V^2}\int d\mathbf{r}d\mathbf{r}'w(\mathbf{r})\bar{\Phi}^{-1}(\mathbf{r}-\mathbf{r}')w(\mathbf{r}')\right]Q^n[iw], \qquad (6)$$

where

$$Z_0 = \frac{1}{n!}\left(\frac{\exp\left(\beta/2N\bar{\Phi}(0)\right)Z'}{\lambda^{3N}(T)}\right)^n$$

can be interpreted as the partition function for an ideal gas of non-interacting polymers and

$$Z' = \int D\mathbf{R}\exp[-\beta U_0(\mathbf{R})] \qquad (7)$$

is the path integral of a free polymer in a zero field with elastic energy

$$U_0[\mathbf{R}] = \frac{k_B T}{4R_{g0}^2}\int_0^1 ds\left|\frac{d\mathbf{R}(s)}{ds}\right|^2.$$

In the latter equation the unperturbed radius of gyration of a chain $R_{g0} = \sqrt{Nb^2/(6)}$. Moreover, in Eq. (6) the partition function of a single polymer, subjected to the field $w(\mathbf{R})$, is given by

$$Q[iw] = \frac{\int D\mathbf{R}\exp\left[-\beta U_0[\mathbf{R}] - iN\int_0^1 ds\, w(\mathbf{R}(s))\right]}{\int D\mathbf{R}\exp[-\beta U_0[\mathbf{R}]]}. \qquad (8)$$

Grand Canonical Ensemble

Basic field-theoretic Representation of Grand Canonical Partition Function

To derive the grand canonical partition function, we use its standard thermodynamic relation to the canonical partition function, given by,

$$\Xi(\mu,V,\beta)=\sum_{n=0}^{\infty}e^{\beta\mu n}Z(n,V,\beta),$$

where μ is the chemical potential and $Z(n,V,\beta)$ is given by Eq. (6). Performing the sum, this provides the field-theoretic representation of the grand canonical partition function,

$$\Xi(\xi,V,\beta)=\gamma_{\bar{\Phi}}\int Dw\exp\left[-S[w]\right],$$

where

$$S[w]=\frac{1}{2\beta V^2}\int d\mathbf{r}d\mathbf{r}'w(\mathbf{r})\bar{\Phi}^{-1}(\mathbf{r}-\mathbf{r}')w(\mathbf{r}')-\xi Q[iw]$$

is the grand canonical action with $Q[iw]$ defined by Eq. (8) and the constant,

$$\gamma_{\bar{\Phi}}=\frac{1}{\sqrt{2}}\prod_{\mathbf{G}}\left(\frac{1}{\pi\beta\bar{\Phi}(\mathbf{G})}\right)^{1/2}.$$

Moreover, the parameter related to the chemical potential is given by,

$$\xi=\frac{\exp\left(\beta\mu+\beta/2N\bar{\Phi}(0)\right)Z'}{\lambda^{3N}(T)},$$

where Z' is provided by Eq. (7).

Mean Field Approximation

A standard approximation strategy for polymer field theories is the mean field (MF) approximation, which consists in replacing the many-body interaction term in the action by a term where all bodies of the system interact with an average effective field. This approach reduces any multi-body problem into an effective one-body problem by assuming that the partition function integral of the model is dominated by a single field configuration. A major benefit of solving problems with the MF approximation, or its numerical implementation commonly referred to as the self-consistent field theory (SCFT), is that it often provides some useful insights into the properties and behavior of complex many-body systems at relatively low computational cost. Successful applications of this approximation strategy can be found for various systems of polymers and complex fluids, like e.g. strongly segregated block copolymers of high molecular weight, highly concentrated neutral polymer solutions or highly concentrated block polyelectrolyte (PE) solutions (Schmid 1998, Matsen 2002, Fredrickson 2002). There are, however, a multitude of cases for which SCFT provides inaccurate or even qualitatively incorrect results (Baeurle 2006a). These comprise neutral polymer or polyelectrolyte solutions in dilute and semidilute concentration regimes, block copolymers near their order-disorder transition, polymer blends near their phase transitions, etc. In such situations the partition function integral defining the field-theoretic model is not entirely dominated by a single MF configuration and field configurations far from it can make important contributions, which require the use of more sophisticated calculation techniques beyond the MF level of approximation.

Higher-order Corrections

One possibility to face the problem is to calculate higher-order corrections to the MF approximation. Tsonchev et al. developed such a strategy including leading (one-loop) order fluctuation corrections, which allowed to gain new insights into the physics of confined PE solutions (Tsonchev 1999). However, in situations where the MF approximation is bad many computationally demanding higher-order corrections to the integral are necessary to get the desired accuracy.

Renormalization Techniques

An alternative theoretical tool to cope with strong fluctuations problems occurring in field theories has been provided in the late 1940s by the concept of renormalization, which has originally been devised to calculate functional integrals arising in quantum field theories (QFT's). In QFT's a standard approximation strategy is to expand the functional integrals in a power series in the coupling constant using perturbation theory. Unfortunately, generally most of the expansion terms turn out to be infinite, rendering such calculations impracticable (Shirkov 2001). A way to remove the infinities from QFT's is to make use of the concept of renormalization (Baeurle 2007). It mainly consists in replacing the bare values of the coupling parameters, like e.g. electric charges or masses, by renormalized coupling parameters and requiring that the physical quantities do not change under this transformation, thereby leading to finite terms in the perturbation expansion. A simple physical picture of the procedure of renormalization can be drawn from the example of a classical electrical charge, Q, inserted into a polarizable medium, such as in an electrolyte solution. At a distance r from the charge due to polarization of the medium, its Coulomb field will effectively depend on a function $Q(r)$, i.e. the effective (renormalized) charge, instead of the bare electrical charge, Q. At the beginning of the 1970s, K.G. Wilson further pioneered the power of renormalization concepts by developing the formalism of renormalization group (RG) theory, to investigate critical phenomena of statistical systems (Wilson 1971).

Renormalization Group Theory

The RG theory makes use of a series of RG transformations, each of which consists of a coarse-graining step followed by a change of scale (Wilson 1974). In case of statistical-mechanical problems the steps are implemented by successively eliminating and rescaling the degrees of freedom in the partition sum or integral that defines the model under consideration. De Gennes used this strategy to establish an analogy between the behavior of the zero-component classical vector model of ferromagnetism near the phase transition and a self-avoiding random walk of a polymer chain of infinite length on a lattice, to calculate the polymer excluded volume exponents (de Gennes 1972). Adapting this concept to field-theoretic functional integrals, implies to study in a systematic way how a field theory model changes while eliminating and rescaling a certain number of degrees of freedom from the partition function integral (Wilson 1974).

Hartree Renormalization

An alternative approach is known as the *Hartree approximation* or *self-consistent one-loop approximation* (Amit 1984). It takes advantage of Gaussian fluctuation corrections to the 0th-order MF contribution, to renormalize the model parameters and extract in a self-consistent way the dominant length scale of the concentration fluctuations in critical concentration regimes.

Tadpole Renormalization

In a more recent work Efimov and Nogovitsin showed that an alternative renormalization technique originating from QFT, based on the concept of *tadpole renormalization*, can be a very effective approach for computing functional integrals arising in statistical mechanics of classical many-particle systems (Efimov 1996). They demonstrated that the main contributions to classical partition function integrals are provided by low-order tadpole-type Feynman diagrams, which account for divergent contributions due to particle self-interaction. The renormalization procedure performed in this approach effects on the self-interaction contribution of a charge (like e.g. an electron or an ion), resulting from the static polarization induced in the vacuum due to the presence of that charge (Baeurle 2007). As evidenced by Efimov and Ganbold in an earlier work (Efimov 1991), the procedure of tadpole renormalization can be employed very effectively to remove the divergences from the action of the basic field-theoretic representation of the partition function and leads to an alternative functional integral representation, called the Gaussian equivalent representation (GER). They showed that the procedure provides functional integrals with significantly ameliorated convergence properties for analytical perturbation calculations. In subsequent works Baeurle et al. developed effective low-cost approximation methods based on the tadpole renormalization procedure, which have shown to deliver useful results for prototypical polymer and PE solutions (Baeurle 2006a, Baeurle 2006b, Baeurle 2007a).

Numerical Simulation

Another possibility is to use Monte Carlo (MC) algorithms and to sample the full partition function integral in field-theoretic formulation. The resulting procedure is then called a polymer field-theoretic simulation. In a recent work, however, Baeurle demonstrated that MC sampling in conjunction with the basic field-theoretic representation is impracticable due to the so-called numerical sign problem (Baeurle 2002). The difficulty is related to the complex and oscillatory nature of the resulting distribution function, which causes a bad statistical convergence of the ensemble averages of the desired thermodynamic and structural quantities. In such cases special analytical and numerical techniques are necessary to accelerate the statistical convergence (Baeurle 2003, Baeurle 2003a, Baeurle 2004).

Mean Field Representation

To make the methodology amenable for computation, Baeurle proposed to shift the contour of integration of the partition function integral through the homogeneous MF solution using Cauchy's integral theorem, providing its so-called *mean-field representation*. This strategy was previously successfully employed by Baer et al. in field-theoretic electronic structure calculations (Baer 1998). Baeurle could demonstrate that this technique provides a significant acceleration of the statistical convergence of the ensemble averages in the MC sampling procedure (Baeurle 2002, Baeurle 2002a).

Gaussian Equivalent Representation

In subsequent works Baeurle et al. (Baeurle 2002, Baeurle 2002a, Baeurle 2003, Baeurle 2003a, Baeurle 2004) applied the concept of tadpole renormalization, leading to the *Gaussian equivalent representation*of the partition function integral, in conjunction with advanced MC techniques in the grand canonical ensemble. They could convincingly demonstrate that this strategy provides a further boost in the statistical convergence of the desired ensemble averages (Baeurle 2002).

Flory–Fox Equation

In polymer chemistry, the Flory–Fox equation is a simple empirical formula that relates molecular weight to the glass transition temperature of a polymer system. The equation was first proposed in 1950 by Paul J. Flory and Thomas G. Fox while at Cornell University. Their work on the subject overturned the previously held theory that the glass transition temperature was the temperature at which viscosity reached a maximum. Instead, they demonstrated that the glass transition temperature is the temperature at which the free space available for molecular motions achieved a minimum value. While its accuracy is usually limited to samples of narrow range molecular weight distributions, it serves as a good starting point for more complex structure-property relationships.

Overview

The Flory–Fox equation relates the number-average molecular weight, M_n, to the glass transition temperature, T_g, as shown below:

$$T_g = T_{g,\infty} - \frac{K}{M_n}$$

where $T_{g,\infty}$ is the maximum glass transition temperature that can be achieved at a theoretical infinite molecular weight and K is an empirical parameter that is related to the free volume present in the polymer sample. It is this concept of "free volume" that is observed by the Flory–Fox equation.

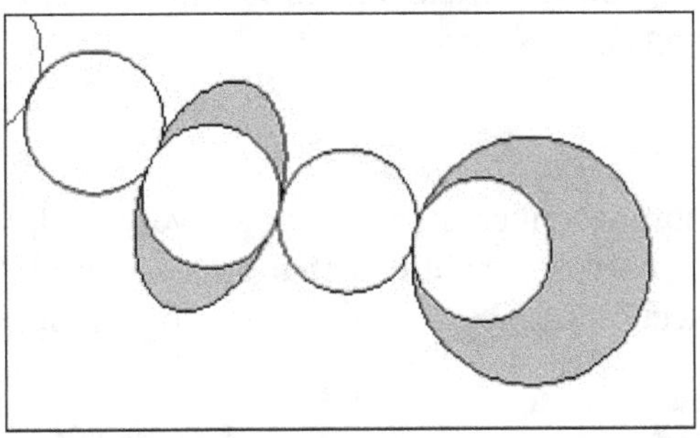

A polymer chain (represented by the white circles) exhibits more free volume (represented by the gray shading) at the ends of the chain than from units within the chain.

Free volume can be most easily understood as a polymer chain's "elbow room" in relation to the other polymer chains surrounding it. The more elbow room a chain has, the easier it is for the chain to move and achieve different physical conformations. Free volume decreases upon cooling from the rubbery state until the glass transition temperature at which point it reaches some critical minimum value and molecular rearrangement is effectively "frozen" out, so the polymer chains lack sufficient free volume to achieve different physical conformations. This ability to achieve different physical conformations is called segmental mobility.

Free volume not only depends on temperature, but also on the number of polymer chain ends present in the system. End chain units exhibit greater free volume than units within the chain because the covalent bonds that make up the polymer are shorter than the intermolecular nearest neighbor distances found at the end of the chain. In other words, end chain units are less dense than the covalently bonded interchain units. This means that a polymer sample with long chain lengths (high molecular weights) will have fewer chain ends per total units and less free volume than a polymer sample consisting of short chains. In short, chain ends can be viewed as an "impurity" when considering the packing of chains, and more impurity results in a lower T_g.

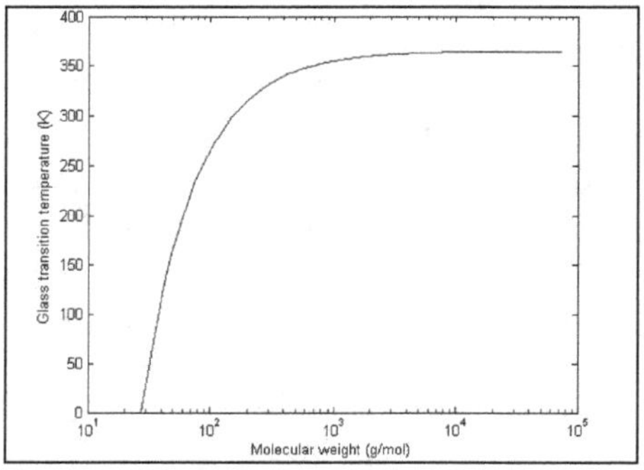

A plot showing the dependence of glass transition temperature (Kelvin) on molecular weight of the polymer chain (g/mol), as predicted by the Flory-Fox equation.

Thus, glass transition temperature is dependent on free volume, which in turn is dependent on the average molecular weight of the polymer sample. This relationship is described by the Flory–Fox equation. Low molecular weight values result in lower glass transition temperatures whereas increasing values of molecular weight result in an asymptotic approach of the glass transition temperature to $T_{g,\infty}$. The figure to the left clearly displays this relationship – as molecular weight increases, the glass transition temperature increases asymptotically to $T_{g,\infty}$ (in this arbitrary case shown in the image, $T_{g,\infty}$ = 365 K).

Alternative Equations

While the Flory–Fox equation describes many polymers very well, it is more reliable for large values of M_n and samples of narrow weight distribution. As a result, other equations have been proposed to provide better accuracy for certain polymers. For example:

$$T_g = T_{g,\infty} - \frac{K}{(M_n M_w)^{\frac{1}{2}}}.$$

This minor modification of the Flory–Fox equation, proposed by Ogawa, replaces the inverse dependence on M_n with the square of the product of the number-average molecular weight, M_n , and weight-average molecular weight, M_w . Additionally, the equation:

$$\frac{1}{T_g} = \frac{1}{T_{g,\infty}} + \frac{K}{T_{g,\infty}^2}\frac{1}{M_n}$$

was proposed by Fox and Loshaek, and has been applied to polystyrene, polymethylmethacrylate, and polyisobutylene, among others.

However, it is important to note that despite the dependence of T_g on molecular weight that the Flory-Fox and related equations describe, molecular weight is not necessarily a practical design parameter for controlling T_g because the range over which it can be changed without altering the physical properties of the polymer due to molecular weight change is small.

The Fox Equation

The Flory–Fox equation serves the purpose of providing a model for how glass transition temperature changes over a given molecular weight range. Another method to modify the glass transition temperature is to add a small amount of low molecular weight diluent, commonly known as a plasticizer, to the polymer. The presence of a low molecular weight additive increases the free volume of the system and subsequently lowers T_g, thus allowing for rubbery properties at lower temperatures. This effect is described by the Fox equation:

$$\frac{1}{T_g} = \frac{w_1}{T_{g,1}} + \frac{w_2}{T_{g,2}}.$$

Where w_1 and w_2 are weight fractions of components 1 and 2, respectively. In general, the accuracy of the Fox equation is very good and it is commonly also applied to predict the glass transition temperature in (miscible) polymer blends and statistical copolymers.

Mayo–Lewis Equation

The Mayo–Lewis equation or copolymer equation in polymer chemistry describes the distribution of monomers in a copolymer: It is named for Frank R. Mayo and Frederick M. Lewis.

Taking into consideration a monomer mix of two components M_1 and M_2 and the four different reactions that can take place at the reactive chain end terminating in either monomer (M^*) with their reaction rate constants k:

$$M_1^* + M_1 \xrightarrow{k_{11}} M_1M_1^*$$

$$M_1^* + M_2 \xrightarrow{k_{12}} M_1M_2^*$$

$$M_2^* + M_2 \xrightarrow{k_{22}} M_2M_2^*$$

$$M_2^* + M_1 \xrightarrow{k_{21}} M_2M_1^*$$

and with reactivity ratios defined as:

$$r_1 = \frac{k_{11}}{k_{12}}$$

$$r_2 = \frac{k_{22}}{k_{21}}$$

the copolymer equation is given as:

$$\frac{d[M_1]}{d[M_2]}=\frac{[M_1](r_1[M_1]+[M_2])}{[M_2]([M_1]+r_2[M_2])}$$

with the concentration of the components given in square brackets. The equation gives the copolymer composition at any instant during the polymerization.

Equation Derivation

Monomer 1 is consumed with reaction rate:

$$\frac{-d[M_1]}{dt}=k_{11}[M_1]\Sigma[M_1^*]+k_{21}[M_1]\Sigma[M_2^*]$$

with $\Sigma[M_x^*]$ the concentration of all the active centers terminating in monomer 1 or 2.

Likewise the rate of disappearance for monomer 2 is:

$$\frac{-d[M_2]}{dt}=k_{12}[M_2]\Sigma[M_1^*]+k_{22}[M_2]\Sigma[M_2^*]$$

Division of both equations yields:

$$\frac{d[M_1]}{d[M_2]}=\frac{[M_1]}{[M_2]}\left(\frac{k_{11}\frac{\Sigma[M_1^*]}{\Sigma[M_2^*]}+k_{21}}{k_{12}\frac{\Sigma[M_1^*]}{\Sigma[M_2^*]}+k_{22}}\right)$$

The ratio of active center concentrations can be found assuming steady state with:

$$\frac{d\,\Sigma[M_1^*]}{dt}=\frac{d\,\Sigma[M_2^*]}{dt}\approx 0$$

meaning that the concentration of active centres remains constant, the rate of formation for active center of monomer 1 is equal to the rate of their destruction or:

$$k_{21}[M_1]\Sigma[M_2^*]=k_{12}[M_2]\Sigma[M_1^*]$$

or

$$\frac{\Sigma[M_1^*]}{\Sigma[M_2^*]}=\frac{k_{21}[M_1]}{k_{12}[M_2]}$$

Substituting into the ratio of monomer consumption rates eliminates the radical concentrations and yields the Mayo-Lewis equation.

Instantaneous Form

It can often be useful to alter the copolymer equation by expressing concentration in terms of mole fractions. Mole fractions of monomers M_1 and M_2 in the feed are defined as f_1 and f_2 where,

$$f_1 = 1 - f_2 = \frac{M_1}{(M_1 + M_2)}$$

Similarly, F represents the mole fraction of each monomer in the copolymer:

$$F_1 = 1 - F_2 = \frac{dM_1}{d(M_1 + M_2)}$$

These equations can be combined with the Mayo-Lewis Equation to give,

$$F_1 = 1 - F_2 = \frac{r_1 f_1^2 + f_1 f_2}{r_1 f_1^2 + 2 f_1 f_2 + r_2 f_2^2}$$

This equation gives the instantaneous copolymer composition. It is important to note that the feed and copolymer compositions can change as polymerization proceeds.

Limiting Cases

Reactivity ratios indicate preference for propagation. Large r indicates a tendency for M_1^* to add M_1 , while small values indicate a tendency for M_1^* to add M_2 . From the definition of reactivity ratios, several special cases can be derived:

- $r_1 = r_2 >> 1$ with both reactivity ratios very high the two monomers only react with themselves and not each other leading to a mixture of two homopolymers.
- $r_1 = r_2 > 1$ with both ratios larger than 1, homopolymerization of component M_1 is favored but in the event of a crosspolymerization by M_2 the chain-end will continue giving rise to block copolymer.
- $r_1 = r_2 \approx 1$ with both ratios around 1, monomer 1 will react as fast with another monomer 1 or monomer 2 and a random copolymer is formed.
- $r_1 = r_2 \approx 0$ with both values approaching 0 the monomers are unable to homopolymerize and only add each other resulting in an alternating polymer.
- $r_1 >> 1 >> r_2$ In the initial stage of the copolymerization monomer 1 is incorporated faster and the copolymer is rich in monomer 1. When this monomer gets depleted, more monomer 2 segments are added. This is called composition drift.

An example case is maleic anhydride and styrene, with reactivity ratios:

- Maleic anhydride (r_1 = 0.01) & styrene (r_2, = 0.02).

Neither of these compounds homopolymerize and instead they react together to give almost exclusively alternating copolymer.

When both $r < 1$, the system has an azeotrope, where feed and copolymer composition are the same.

Calculation of Reactivity Ratios

Calculation of reactivity ratios generally involves carrying out several polymerizations at varying monomer ratios. The copolymer composition can be analysed with methods such as Proton nuclear magnetic resonance, Carbon-13 nuclear magnetic resonance, or Fourier transform infrared spectroscopy. The polymerizations are also carried out at low conversions, so monomer concentrations can be assumed to be constant. With all the other parameters in the copolymer equation known, r_1 and r_2 can be found.

Curve Fitting

One of the simplest methods for finding reactivity ratios is plotting the copolymer equation and using least squares analysis to find the r_1 , r_2 pair that gives the best fit curve.

Mayo-Lewis Method

The Mayo-Lewis method uses a form of the copolymer equation relating r_1 to r_2 :

$$r_2 = \frac{f_1}{f_2}\left[\frac{F_2}{F_1}(1+\frac{f_1 r_1}{f_2})-1\right]$$

For each different monomer composition, a line is generated using arbitrary r_1 values. The intersection of these lines is the r_1 , r_2 for the system. More frequently, the lines do not intersect and the area in which most line intersect can be given as a range of r_1 , and r_2 values.

Fineman-Ross Method

Fineman and Ross rearranged the copolymer equation into a linear form:

$$G = Hr_1 - r_2$$

where $G = \frac{f_1(2F_1 - 1)}{(1-f_1)F_1}$ and $H = \left[\frac{f_1^2(1-F_1)}{(1-f_1)^2 F_1}\right]$.

Thus, a plot of H versus G yields a straight line with slope r_1 and intercept $-r_2$.

Kelen Tudos Method

The Fineman-Ross method can be biased towards points at low or high monomer concentration, so Kelen and Tudos introduced and arbitrary constant,

$$\alpha = (H_{min} H_{max})^{0.5}$$

where H_{min} and H_{max} are the highest and lowest values of H from the Fineman-Ross method. The data can be plotted in a linear form,

$$\eta = \left[r_1 + \frac{r_2}{\alpha}\right]\mu - \frac{r_2}{\alpha}$$

where $\eta = G/(\alpha + H)$ and $\mu = H/(\alpha + H)$. Plotting η against μ yields a straight line that gives $-r_2/\alpha$ when $\mu = 0$ and r_1 when $\mu = 1$. This distributes the data more symmetrically and can yield better results.

Q-e Scheme

A semi-empirical method for the determination of reactivity ratios is called the Q-e scheme. This involves using two parameters for each monomer, Q and e. The reaction of M_1 radical with M_2 monomer is written as:

$$k_{12} = P_1 Q_2 exp(-e_1 e_2)$$

while the reaction of M_1 radical with M_1 monomer is written as:

$$k_{11} = P_1 Q_1 exp(-e_1 e_1)$$

Where Q is the measure of reactivity of monomer via resonance stabilization, and e is the measure of polarity of monomer (molecule or radical) via the effect of functional groups on vinyl groups. Using these definitions, r_1 and r_2 can be found by the ratio of the terms. An advantage of this system is that reactivity ratios can be found using tabulated Q-e values of monomers regardless or what the monomer pair is in the system.

References

- Franks, F. "Water Solubility and Sensitivity-Hydration Effects.". In Finch, C. Chemistry and Technology of Water-Soluble Polymers. Springer. pp. 157–178. ISBN 978-1-4757-9661-2
- Muthukumar, M (2004). "Nucleation in Polymer Crystallization". Advances in Chemical Physics. 128. ISBN 0-471-44528-2
- Fox, T.G.; Loshaek, S. (1955), "Influence of molecular weight and degree of crosslinking on the specific volume and glass temperature of polymers", Journal of Polymer Science, 371: 15, doi:10.1002/pol.1955.120158006
- Young, Robert J. Introduction to polymers ([Reprinted with additional material] ed.). London: Chapman and Hall. ISBN 0-412-22170-5
- Amit, D.J. "Field theory, the renormalization group, and critical phenomena". Singapore, World Scientific. ISBN 9812561196
- Burchard, W. "Solution Thermodynamics of Non-Ionic Water Soluble Polymers.". In Finch, C. Chemistry and Technology of Water-Soluble Polymers. Springer. pp. 125–142. ISBN 978-1-4757-9661-2

7

Types of Polymerization

The chemical reaction in which monomer molecules react together to form polymer chains is known as polymerization. In radical polymerization method of polymerization, the polymer forms by the successive addition of free radical blocks. This chapter explains briefly various types of polymerization like step-growth polymerization, chain-growth polymerization, radical polymerization and copolymerization.

Polymerization is a process in which relatively small molecules, called monomers, combine chemically to produce a very large chainlike or network molecule, called a polymer.

The monomer molecules may be all alike, or they may represent two, three, or more different compounds. Usually at least 100 monomer molecules must be combined to make a product that has certain unique physical properties—such as elasticity, high tensile strength, or the ability to form fibres—that differentiate polymers from substances composed of smaller and simpler molecules; often, many thousands of monomer units are incorporated in a single molecule of a polymer. The formation of stable covalent chemical bonds between the monomers sets polymerization apart from other processes, such as crystallization, in which large numbers of molecules aggregate under the influence of weak intermolecular forces.

Two classes of polymerization usually are distinguished. In condensation polymerization, each step of the process is accompanied by the formation of a molecule of some simple compound, often water. In addition polymerization, monomers react to form a polymer without the formation of by-products. Addition polymerizations usually are carried out in the presence of catalysts, which in certain cases exert control over structural details that have important effects on the properties of the polymer.

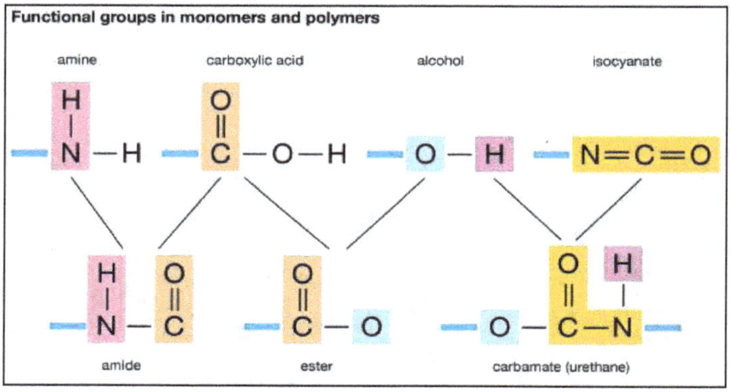

Functional group: monomers and polymers Functional groups in monomers and polymers.

Linear polymers, which are composed of chainlike molecules, may be viscous liquids or solids with varying degrees of crystallinity; a number of them can be dissolved in certain liquids, and they soften or melt upon heating. Cross-linked polymers, in which the molecular structure is a network, are thermosetting resins (i.e., they form under the influence of heat but, once formed, do not melt or soften upon reheating) that do not dissolve in solvents. Both linear and cross-linked polymers can be made by either addition or condensation polymerization.

Classification

The polymerization reactions are classified as:

- Radical Polymerization
- Ionic Polymerization:
    - Anionic polymerization
    - Cationic polymerization

Radical Polymerization

In order to produce radical polymerization it´s necessary that initiator monomer or hardener, which activates and causes the reaction, contains free radicals, namely unpaired electrons which react with the resin monomer in order to form the polymer.

Radical is defined as chemical material extremely unstable therefore with a high reactive power due it has unpaired electrons.

The steps that occur in the radical polymerization are:

Initiation of the reaction - free radical is formed by the action of chemical, thermal, electrochemical or photochemical energy.

- Formation of radical-R --- RAD> RAD * + * R
- Chain initiation RAD * + A ---> RAD-A *

Chain growth

- RAD-A * + A ---> RAD-A-A *
- RAD-A-A * + A ---> RAD-A-A-A *

Chain termination - There are several ways to end the reaction by either:

- Combine 2 radicals RAD-AAA-AAA * + * --- RAD> RAD-AAAAAA-RAD
- Using inhibitors / regulators – external agents of the polymerization which react with the radical making terminate the reaction.

One of the main disadvantages of radical polymerization is that it can not control the molecular weight and size of the final polymer, because each reaction terminates in an undefined way.

The polymers that are produced by radical polymerization are strongly temperature dependent, if temperature increase will speed up the polymerization reactions causing:

- Shorter chains of the polymers and decreasing their mechanical properties (strength and elongation).
- Reduction of working time, pot life and curing.

Ionic Polymerization

Ionic polymerization operation is similar to the radical polymerization; the radical in this case is an ion (atom or molecule) in which one of its parts is charged positively or negatively by the absence or presence of electrons.

Features:

- Requires less activation energy than the radical polymerization.
- It is not as temperature dependent.
- The chain termination occurs only for the use of inhibitors, regulators or other agents that stop the reaction.

Depending on the charge of the ion, ionic polymerization is classified into:

- Anionic polymerization
- Cationic Polymerization

Anionic polymerization - The ion is an atom or molecule with excess of electrons - negatively charged.

ION-+ A -> ION-A-

ION-A-+ A -> ION-A-A-

ION-A-A-+ A -> ION-A-A-A-

Cationic Polymerization - The ion is an atom or molecule with lack of electrons - positively charged.

ION + + A -> ION-A +

ION-A + + A -> ION-A-A +

ION-A-A + + A -> ION-A-A-A +

Adhesives Cure by Polymerization:

- Radical polymerization
    - MMA
    - Anaerobic
    - Unsaturated polyesters
    - Acrylic curing by ultraviolet
- Ionic Polymerization:
    - Anionic - Cyanoacrylates
    - Cationic - Radiation curing Epoxy

Radical Polymerization

Radical polymerization belongs to the most important reactions running by chain mechanism. Mechanism of polymer chain creation can be described by elementary actions: initiation, propagation, termination and transfer reactions. Succession of actions running at construction of one macromolecule is very quick (around one second). However, great number of macromolecules is produced simultaneously in the system, which is why all types of partial reaction occur simultaneously.

Initiation

The initiation reaction produces free radicals. There are several ways to do this:

- Chemical initiation The decomposition of the initiator (e.g. AIBN) forms free radicals:

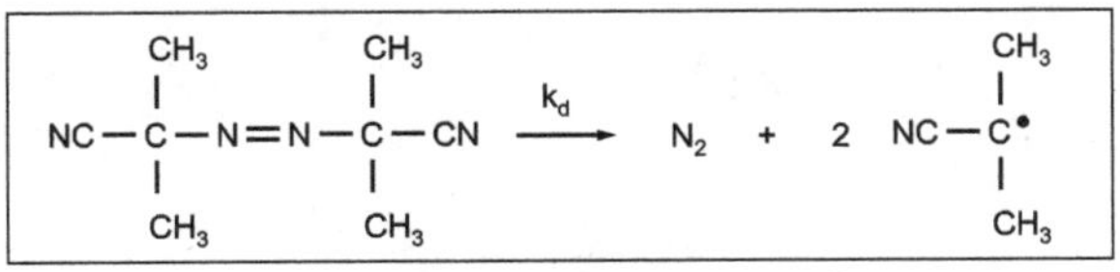

$$I_2 \xrightarrow{k_d} I^\bullet + I^\bullet \quad (r_d = k_d I_2)$$

$$I^\bullet + M \xrightarrow{k_I} R_1^\bullet$$

$$\frac{dI^\bullet}{dt} = 2f\, k_d I_2 - M \approx 0$$

$$\Rightarrow k_I I^\bullet M = 2f k_d I_2 \equiv R_I$$

where f is the initiator efficiency, typically f = [0.5, 1]. Note that in order to ensure a continuous production of radicals all over the process, $1/k_d$ should be larger than the characteristic time of the polymerization reaction. Examples of the decomposition characteristic time, τ_d for some commercial initiators are:

	τ_d	T
Acetyl peroxide	2 h	80° C
Cumyl peroxide	12 h	110° C
t- Butyl hydroperoxide	45 h	150° C

Since this is a first order process, $\tau_d = 1/k_d$.

1. Thermal Initiation: Thermal Decomposition of the Monomer (e.g. styrene)

   This represents a danger, for example during monomer transportation, since it may lead to undesired polymerization of the monomer. For this reason, inhibitors (scavengers of radicals) are usually added to the monomers before storage. This causes the occurrence of a non reproducible induction period when such monomers are polymerized.

2. Initiation by Radiation

   The decomposition of the initiator is caused by light or another source of radiation. Since this method is quite expensive, it is only applied to polymerization systems operating at very low temperatures.

Propagation

Propagation is the addition of a monomer molecule to a radical chain.

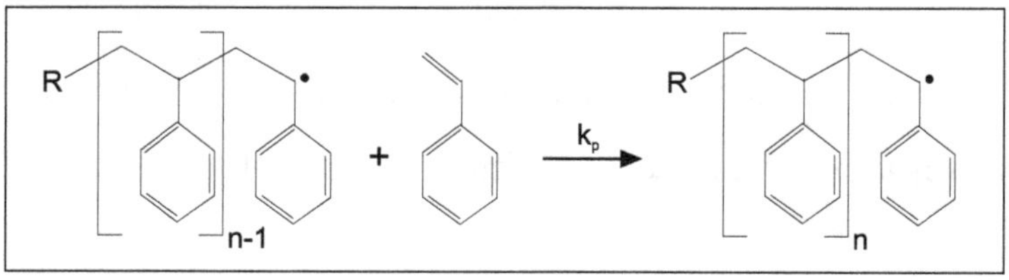

$$R_n^\bullet + M \xrightarrow{k_d} R_{n+1}^\bullet \quad \left(r_p = k_p R_n^\bullet M\right)$$

Chain Transfer

- Chain transfer to monomer

$$R_n^\bullet + M \xrightarrow{k_{fm}} P_n + R_1^\bullet \quad \left(r = k_{fm} R_n^\bullet M\right)$$

The reactants are the same as for the propagation reaction, but the activation energy is much larger. Accordingly, k_{fm} is usually at least 103 times smaller than k_p. This reaction

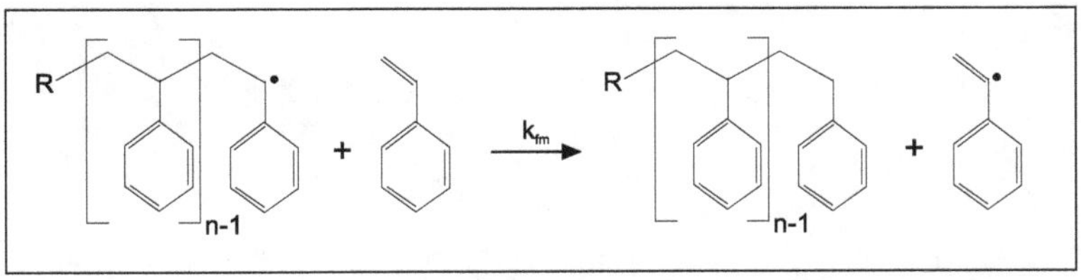

leads to the formation of a polymer chain with a terminal double bond. This can induce chain branching through the terminal double bond propagation reaction.

- Chain transfer to chain transfer agent

$$R_n^{\bullet} + P_m \xrightarrow{k_{fs}} P_n + R_1^{\bullet} \quad \left(r = k_{fs}\, R_n^{\bullet} S\right)$$

A chain transfer agent, S is a molecule containing a weak bond that can be broken to lead to radical transfer, similarly as in the case of monomer above (e.g. CCl_4, CBr_4, mercaptans).

- Chain transfer to polymer

$$R_n^{\bullet} + P_m \xrightarrow{k_{fP}} P_n + R_m^{\bullet} \quad \left(r = k_{fp}\, R_n^{\bullet} \left(mP_m\right)\right)$$

In this reaction the growing radical chain, R˙ n extracts a hydrogen from the dead chain, Pm. Since this extraction can occur on any of the m monomer units along the chain, the rate of this reaction is proportional to the length of P_m.

General observations on the role of chain transfer reactions:

- The concentration of radicals is not affected and therefore the rate of monomer consumption is also unchanged.
- The growth of polymer chains is stopped and therefore shorter chains are produced.
- Each transfer event leaves a different end-group on the chain that can be detected (NMR, titration) so as to identify and quantify the corresponding chain transfer reaction.
- Nonlinear (branched) polymer chains are produced: directly by chain transfer to polymer or indirectly through the propagation of the terminal double bond left by a chain transfer to monomer event.

Bimolecular Termination

Bimolecular termination occurs according to two different mechanisms: termination by combination and termination by disproportionation. Their relative importance depends upon the specific polymerization system.

- Termination by combination

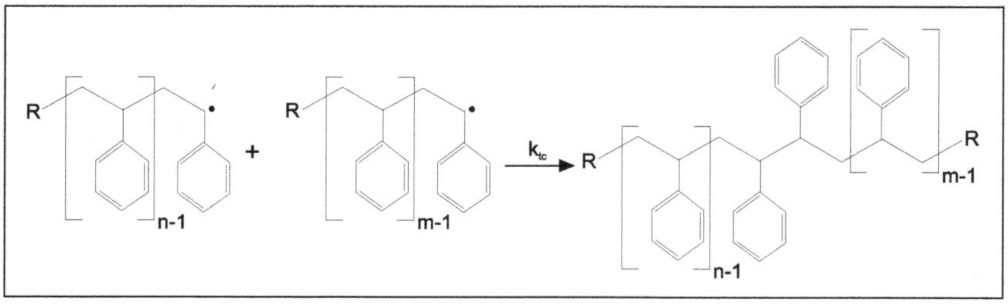

$$R_n^{\bullet} + R_m^{\bullet} \xrightarrow{k_{tc}} P_{n+m} \quad \left(r = k_{tc} R_n^{\bullet} R_m^{\bullet}\right)$$

This reaction results in an increase of the chain length.

- Termination by disproportionation

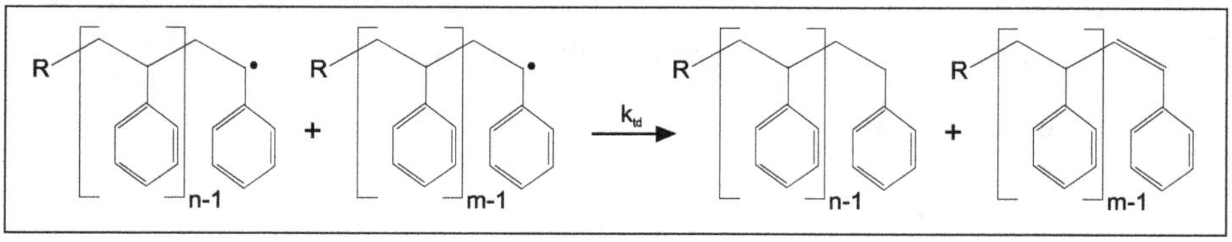

$$R_n^{\bullet} + R_m^{\bullet} \xrightarrow{k_{td}} P_n + P_m \quad \left(r = k_{td} R_n^{\bullet} R_m^{\bullet}\right)$$

The chain length remains constant during the termination reaction. Note that also in this case chains with terminal double bond are produced, which can therefore lead to the occurrence of branching.

Methods

There are four industrial methods of radical polymerization:

- *Bulk polymerization:* reaction mixture contains only initiator and monomer, no solvent.
- *Solution polymerization:* reaction mixture contains solvent, initiator, and monomer.
- *Suspension polymerization:* reaction mixture contains an aqueous phase, water-insoluble monomer, and initiator soluble in the monomer droplets (both the monomer and the initiator are hydrophobic).
- *Emulsion polymerization:* similar to suspension polymerization except that the initiator is soluble in the aqueous phase rather than in the monomer droplets (the monomer is hydrophobic, and the initiator is hydrophilic). An emulsifying agent is also needed.

Other methods of radical polymerization include the following:

- *Template polymerization*: In this process, polymer chains are allowed to grow along template macromolecules for the greater part of their lifetime. A well-chosen template can affect the rate of polymerization as well as the molar mass and microstructure of the daughter polymer. The molar mass of a daughter polymer can be up to 70 times greater than those of polymers produced in the absence of the template and can be higher in molar mass than the templates themselves. This is because of retardation of the termination for template-associated radicals and by hopping of a radical to the neighboring template after reaching the end of a template polymer.
- *Plasma polymerization*: The polymerization is initiated with plasma. A variety of organic molecules including alkenes, alkynes, and alkanes undergo polymerization to high molecular weight products under these conditions. The propagation mechanisms appear to involve both ionic and radical species. Plasma polymerization offers a potentially unique method of forming thin polymer films for uses such as thin-film capacitors, antireflection coatings, and various types of thin membranes.
- *Sonication*: The polymerization is initiated by high-intensity ultrasound. Polymerization to high molecular weight polymer is observed but the conversions are low (<15%). The polymerization is self-limiting because of the high viscosity produced even at low conversion. High viscosity hinders cavitation and radical production.

Reversible Deactivation Radical Polymerization

Also known as living radical polymerization, controlled radical polymerization, reversible deactivation radical polymerization (RDRP) relies on completely pure reactions, preventing termination caused by impurities. Because these polymerizations stop only when there is no more monomer, polymerization can continue upon the addition of more monomer. Block copolymers can be made this way. RDRP allows for control of molecular weight and dispersity. However, this is very difficult to achieve and instead a pseudo-living polymerization occurs with only partial control of molecular weight and dispersity. ATRP and RAFT are the main types of complete radical polymerization.

- *Atom Transfer Radical Polymerization (ATRP):* based on the formation of a carbon-carbon bond by atom transfer radical addition. This method, independently discovered in 1995 by Mitsuo Sawamoto and by Jin-Shan Wang and Krzysztof Matyjaszewski, requires reversible activation of a dormant species (such as an alkyl halide) and a transition metal halide catalyst (to activate dormant species).
- *Reversible Addition-Fragmentation Chain Transfer Polymerization (RAFT):* requires a compound that can act as a reversible chain transfer agent, such as dithio compounds.

- *Stable Free Radical Polymerization (SFRP)*: used to synthesize linear or branched polymers with narrow molecular weight distributions and reactive end groups on each polymer chain. The process has also been used to create block co-polymers with unique properties. Conversion rates are about 100% using this process but require temperatures of about 135 °C. This process is most commonly used with acrylates, styrenes, and dienes. The reaction scheme in Figure illustrates the SFRP process.

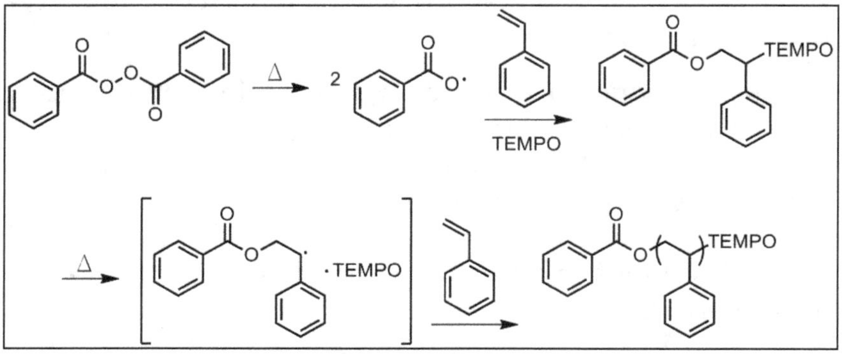

Figure: Reaction scheme for SFRP.

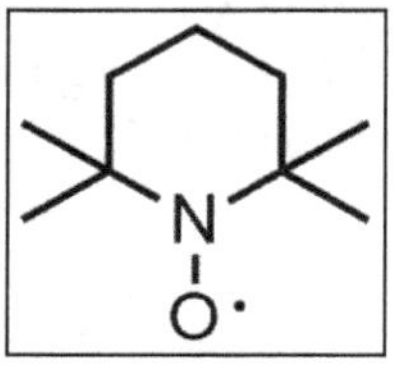

Figure: TEMPO molecule used to functionalize the chain ends.

Because the chain end is functionalized with the TEMPO molecule Figure, premature termination by coupling is reduced. As with all living polymerizations, the polymer chain grows until all of the monomer is consumed.

Kinetics

In typical chain growth polymerizations, the reaction rates for initiation, propagation and termination can be described as follows:

$$v_i = \mathrm{d}[M\cdot]/\mathrm{d}t = 2k_d f[I]$$

$$v_p = k_p[M][M\cdot]$$

$$v_t = -\mathrm{d}[M\cdot]/\mathrm{d}t = 2k_t[M\cdot]^2$$

where *f* is the efficiency of the initiator and k_d, k_p, and k_t are the constants for initiator dissociation, chain propagation and termination, respectively. [I] [M] and [M•] are the concentrations of the initiator, monomer and the active growing chain.

Under the steady state approximation, the concentration of the active growing chains remains constant, i.e. the rates of initiation and of termination are equal. The concentration of active chain can be derived and expressed in terms of the other known species in the system,

$$[M\cdot]=\left(\frac{k_d[I]f}{k_t}\right)^{1/2}$$

In this case, the rate of chain propagation can be further described using a function of the initiator and monomer concentrations,

$$\upsilon_p=k_p\left(\frac{fk_d}{k_t}\right)^{1/2}[I]^{1/2}[M]$$

The kinetic chain length v is a measure of the average number of monomer units reacting with an active center during its lifetime and is related to the molecular weight through the mechanism of the termination. Without chain transfer, the kinetic chain length is only a function of propagation rate and initiation rate,

$$\nu=\frac{\upsilon_p}{\upsilon_i}=\frac{k_p[M][M\cdot]}{2fk_d[I]}=\frac{k_p[M]}{2(fk_dk_t[I])^{1/2}}$$

Assuming no chain transfer effect occurs in the reaction, the number average degree of polymerization P_n can be correlated with the kinetic chain length. In the case of termination by disproportionation, one polymer molecule is produced per every kinetic chain:

$$x_n=\nu$$

Termination by combination leads to one polymer molecule per two kinetic chains:

$$x_n=2\nu$$

Any mixture of both these mechanisms can be described by using the value δ, the contribution of disproportionation to the overall termination process:

$$x_n=\frac{2}{1+\delta}\nu$$

If chain transfer is considered, the kinetic chain length is not affected by the transfer process because the growing free-radical center generated by the initiation step stays alive after any chain transfer event, although multiple polymer chains are produced. However, the number average degree of polymerization decreases as the chain transfers, since the growing chains are terminated by the chain transfer events. Taking into account the chain transfer reaction towards solvent *S*, initiator *I*, polymer *P*, and added chain transfer agent *T*. The equation of P_n will be modified as follows:

$$\frac{1}{x_n}=\frac{2k_{t,d}+k_{t,c}}{k_p{}^2[M]^2}\upsilon_p+C_M+C_S\frac{[S]}{[M]}+C_I\frac{[I]}{[M]}+C_P\frac{[P]}{[M]}+C_T\frac{[T]}{[M]}$$

It is usual to define chain transfer constants C for the different molecules,

$$C_M=\frac{k_{tr}^M}{k_p},\ C_S=\frac{k_{tr}^S}{k_p},\ C_I=\frac{k_{tr}^I}{k_p},\ C_P=\frac{k_{tr}^P}{k_p},\ C_T=\frac{k_{tr}^T}{k_p},$$

Thermodynamics

In chain growth polymerization, the position of the equilibrium between polymer and monomers can be determined by the thermodynamics of the polymerization. The Gibbs free energy (ΔG_p) of the polymerization is commonly used to quantify the tendency of a polymeric reaction. The polymerization will be favored if $\Delta G_p < 0$; if $\Delta G_p > 0$, the polymer will undergo depolymerization. According to the thermodynamic equation $\Delta G = \Delta H - T\Delta S$, a negative enthalpy and an increasing entropy will shift the equilibrium towards polymerization.

In general, the polymerization is an exothermic process, i.e. negative enthalpy change, since addition of a monomer to the growing polymer chain involves the conversion of π bonds into σ bonds, or a ring–opening reaction that releases the ring tension in a cyclic monomer. Meanwhile, during polymerization, a large amount of small molecules are associated, losing rotation and translational degrees of freedom. As a result, the entropy decreases in the system, $\Delta S_p < 0$ for nearly all polymerization processes. Since depolymerization is almost always entropically favored, the ΔH_p must then be sufficiently negative to compensate for the unfavorable entropic term. Only then will polymerization be thermodynamically favored by the resulting negative ΔG_p.

In practice, polymerization is favored at low temperatures: $T\Delta S_p$ is small. Depolymerization is favored at high temperatures: $T\Delta S_p$ is large. As the temperature increases, ΔG_p become less negative. At a certain temperature, the polymerization reaches equilibrium (rate of polymerization = rate of depolymerization). This temperature is called the ceiling temperature (T_c). $\Delta G_p = 0$.

Stereochemistry

The stereochemistry of polymerization is concerned with the difference in atom connectivity and spatial orientation in polymers that has the same chemical composition. Staudinger studied the stereoisomerism in chain polymerization of vinyl monomers in late 1920s, and it took another two decades for people to fully appreciate the idea that each of the propagation steps in the polymer growth could give rise to stereoisomerism.

The major milestone in the stereochemistry was established by Ziegler and Natta and their coworkers in 1950s, as they developed metal based catalyst to synthesize stereoregular polymers. The reason why the stereochemistry of the polymer is of particular interest is because the physical behavior of a polymer depends not only on the general chemical composition but also on the more subtle differences in microstructure. Atactic polymers consist of a random arrangement of stereochemistry and are amorphous (noncrystalline), soft materials with lower physical strength. The corresponding isotactic (like substituents all on the same side) and syndiotactic (like substituents of alternate repeating units on the same side) polymers are usually obtained as highly crystalline materials. It is easier for the stereoregular polymers to pack into a crystal lattice since they are more ordered and the resulting crystallinity leads to higher physical strength and increased solvent and chemical resistance as well as differences in other properties that depend on crystallinity. The prime example of the industrial utility of stereoregular polymers is polypropene. Isotactic polypropene is a high-melting (165 °C), strong, crystalline polymer, which is used as both a plastic and fiber. Atactic polypropene is an amorphous material with an oily to waxy soft appearance that finds use in asphalt blends and formulations for lubricants, sealants, and adhesives, but the volumes are minuscule compared to that of isotactic polypropene.

When a monomer adds to a radical chain end, there are two factors to consider regarding its stereochemistry: 1) the interaction between the terminal chain carbon and the approaching monomer molecule and 2) the configuration of the penultimate repeating unit in the polymer chain. The terminal carbon atom has *sp*² hybridization and is planar. Consider the polymerization of the monomer CH_2=CXY. There are two ways that a monomer molecule can approach the terminal carbon: the mirror approach (with like substituents on the same side) or the non-mirror approach (like substituents on opposite sides). If free rotation does not occur before the next monomer adds, the mirror approach will always lead to an isotactic polymer and the non-mirror approach will always lead to a syndiotactic polymer (Figure).

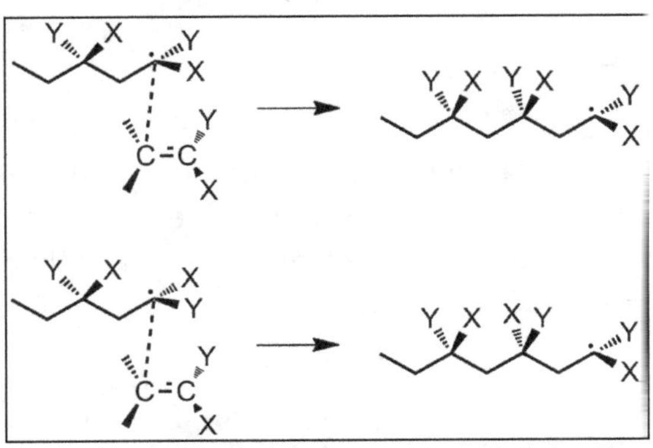

Figure: (Top) formation of isotactic polymer; (bottom) formation of syndiotactic polymer.

However, if interactions between the substituents of the penultimate repeating unit and the terminal carbon atom are significant, then conformational factors could cause the monomer to add to the polymer in a way that minimizes steric or electrostatic interaction Figure.

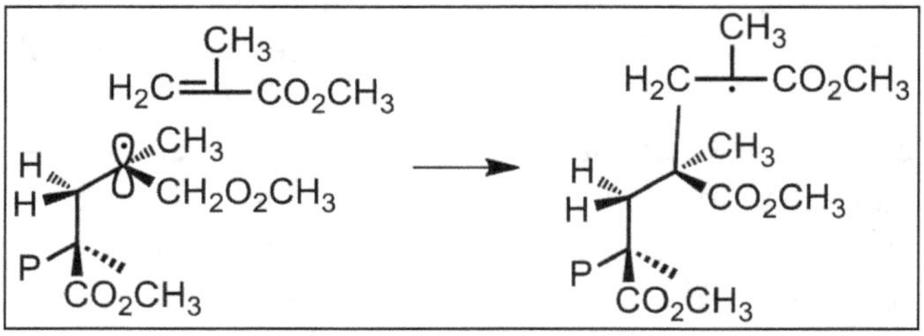

Figure: Penultimate unit interactions cause monomer to add in a way that minimizes steric hindrance between substituent groups. (P represents polymer chain.)

Reactivity

Traditionally, the reactivity of monomers and radicals are assessed by the means of copolymerization data. The *Q–e* scheme, the most widely used tool for the semi-quantitative prediction of monomer reactivity ratios, was first proposed by Alfrey and Price in 1947. The scheme takes into account the intrinsic thermodynamic stability and polar effects in the transition state. A given radical M_i^o and a monomer M_j are considered to have intrinsic reactivities P_i and Q_j, respectively. The polar effects in the transition state, the supposed permanent electric charge carried by that entity (radical or molecule), is quantified by the factor *e*, which is a constant for a given monomer, and has the same value for the radical derived from that specific monomer. For addition of monomer 2 to a growing polymer chain whose active end is the radical of monomer 1, the rate constant, k_{12}, is postulated to be related to the four relevant reactivity parameters by:

$$k_{12} = P_1 Q_2 \exp(-e_1 e_2)$$

The monomer reactivity ratio for the addition of monomers 1 and 2 to this chain is given by:

$$r_1 = \frac{k_{11}}{k_{12}} = \frac{Q_1}{Q_2} \exp(-e_1(e_1 - e_2))$$

For the copolymerization of a given pair of monomers, the two experimental reactivity ratios r_1 and r_2 permit the evaluation of (Q_1/Q_2) and $(e_1 - e_2)$. Values for each monomer can then be assigned relative to a reference monomer, usually chosen as styrene with the arbitrary values Q = 1.0 and e = −0.8.

Applications

Free radical polymerization has found applications including the manufacture of polystyrene, thermoplastic block copolymer elastomers, cardiovascular stents, chemical surfactants and lubricants. Block copolymers are used for a wide variety of applications including adhesives, footwear and toys.

Free radical polymerization has uses in research as well, such as in the functionalization of carbon nanotubes. CNTs intrinsic electronic properties lead them to form large aggregates in solution, precluding useful applications. Adding small chemical groups to the walls of CNT can eliminate this propensity and tune the response to the surrounding environment. The use of polymers instead of smaller molecules can modify CNT properties (and conversely, nanotubes can modify polymer mechanical and electronic properties). For example, researchers coated carbon nanotubes with polystyrene by first polymerizing polystyrene via chain radical polymerization and subsequently mixing it at 130 °C with carbon nanotubes to generate radicals and graft them onto the walls of carbon nanotubes Figure. Chain growth polymerization ("grafting to") synthesizes a polymer with predetermined properties. Purification of the polymer can be used to obtain a more uniform length distribution before grafting. Conversely, "grafting from", with radical polymerization techniques such as atom transfer radical polymerization (ATRP) or nitroxide-mediated polymerization (NMP), allows rapid growth of high molecular weight polymers.

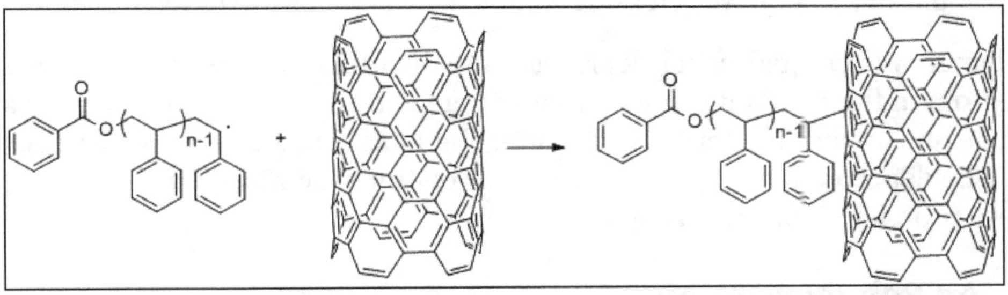

Figure: Grafting of a polystyrene free radical onto a single-walled carbon nanotube.

Radical polymerization also aids synthesis of nanocomposite hydrogels. These gels are made of water-swellable nano-scale clay (especially those classed as smectites) enveloped by a network polymer. They are often biocompatible and have mechanical properties (such as flexibility and strength) that promise applications such as synthetic tissue. Synthesis involves free radical polymerization. The general synthesis procedure is depicted in Figure. Clay is dispersed in water, where it forms very small, porous plates. Next the initiator and a catalyst are added, followed by adding the organic monomer, generally an acrylamide or acrylamide derivative. The initiator is chosen to have stronger interaction with the clay than the organic monomer, so it preferentially adsorbs to the clay surface. The mixture and water solvent is heated to initiate polymerization. Polymers grow off the initiators that are in turn bound to the clay. Due to recombination and disproportionation reactions, growing polymer chains bind to one another, forming a strong, cross-linked network polymer, with clay particles acting as branching points for multiple polymer chain segments. Free radical polymerization used in this context allows the synthesis of polymers from a wide variety of substrates (the chemistries of suitable clays vary). Termination reactions unique to chain growth polymerization produce a material with flexibility, mechanical strength and biocompatibility.

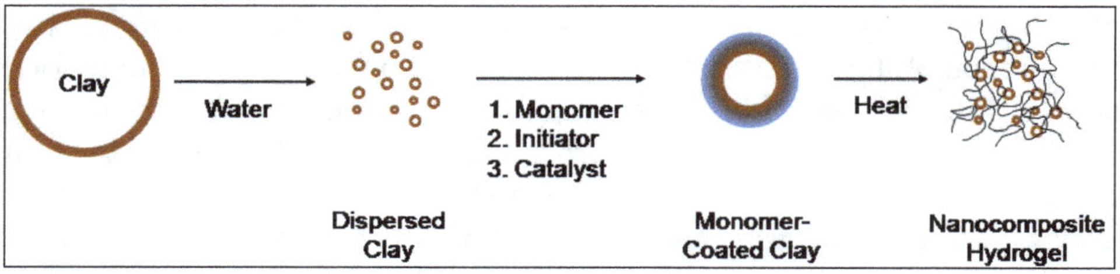

Figure: General synthesis procedure for a nanocomposite hydrogel.

Electronics

The radical polymer glass PTMA is about 10 times more electrically conductive than common semiconducting polymers. PTMA is in a class of electrically active polymers that could find use in transparent solar cells, antistatic and antiglare coatings for mobile phone displays, antistatic coverings for aircraft to protect against lightning strikes, flexible flash drives, and thermoelectric devices, which convert electricity into heat and the reverse. Widespread practical applications require increasing conductivity another 100 to 1,000 times.

The polymer was created using deprotection, which involves replacing a specific hydrogen atom in the pendant group with an oxygen atom. The resulting oxygen atom in PTMA has one unpaired electron in its outer shell, making it amenable to transporting charge. The deprotection step can lead to four distinct chemical functionalities, two of which are promising for increasing conductivity.

Emulsion Polymerization

Emulsion polymerization is a type of radical polymerization. Emulsion polymerization is one of the most important methods for the polymerization of a large number of monomers, like vinyl acetate, vinyl chloride, chloroprene, acrylamide, acrylates, and methacrylates. It is also used for the production of various copolymers, like acrylonitrile-butadiene-styrene (ABS).

Emulsion polymerization has several advantages over other polymerization techniques; for example, it is more rapid than bulk or solution polymerization at the same temperature, the conversion is essentially 100 percent, and the average molecular weight is usually (much) higher than at the same polymerization rate in bulk or solution polymerization. Also, heat dissipation and viscosity control are much less problematic than in bulk polymerization.

In general, an emulsion polymerization system consists of a dispersing medium, monomer, emulsifier, initiator and, if necessary, modifiers. Water[1] is normally the continuous phase in which the various components are dispersed by the emulsifiers. The monomers are only slightly soluble in water. They form droplets that are suspended and stabilized by the emulsifiers, that is, the emulsifier molecules associate and form micelles that surround small amounts of monomer. The remaining monomer is dispersed in small droplets.

Common emulsifiers are anionic and nonionic surfactants. A combination of both types will often improve the stability of the dispersed droplets. Typical anionic emulsifiers are sodium, potassium, or ammonium salts of fatty acids and C_{12} - C_{16}alkyl sulfates. Typical nonionic surfactants are poly(ethylene oxide), poly(vinyl alcohol) and hydroxyethyl cellulose.

The first hypothesis of the mechanism of emulsion polymerization was proposed by-*Harkins*[2]; The process can be divided in three distinctive stages:

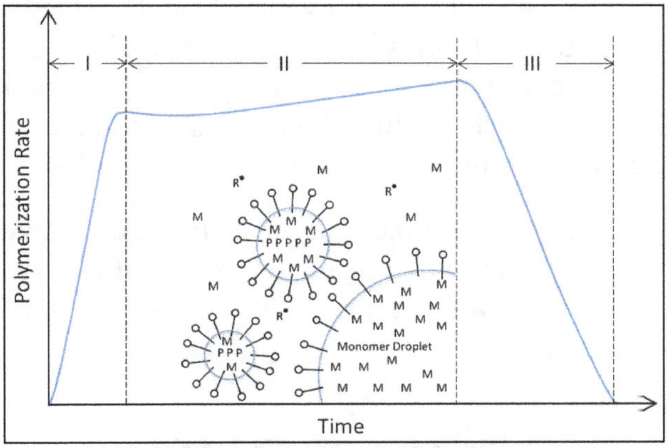

Stages of an emulsion polymerization.

STAGE I

At the beginning, the mixture consists solely of the continuous water phase with dispersed surfactant micelles[3] and emulsified monomer droplets of 1-10 microns. Most monomer is localized in these droplets and some is dissolved in the micelles and a little in the water phase. The micelles are in a dynamic equilibrium with the dissolved emulsifier molecules. Nucleation stops when the surface area becomes large enough to absorb all of the emulsifier molecules. The polymerization starts when initiator is added. A common initiator is (water soluble) potassium persulfate. It decomposes in the water and forms negatively charged sulfate radicals:

$S_2O_8^- + \text{Heat} \rightarrow 2\ SO_{4\cdot}^-$

These radicals react with the in the water dissolved monomers and form soap-type free radicals.

$SO_{4\cdot}^- + (n+1)\ M \rightarrow SO_4^-\text{–}(CH_2\text{-}CX_2)_n\text{–}CH_2\text{-}CX_2^{\cdot}$

They either associate with the dissolved emulsifier molecules to micelles or they migrate into existing micelle droplets. During the first stage of the polymerization, monomers continuously migrate from the large monomer droplets through the water phase into the micelles and are added to the growing polymer chains. At the same time, new particles are formed by the initiators. Some polymerization takes also place in the water phase due to the solubility of the monomers and initiators, whereas the monomer droplets do not provide loci for polymerization because the negatively charged surfactant molecules that surround these droplets are virtually impossible to penetrate by the also negatively charged initiator molecules.

STAGE II

The number of polymer particles and the rate of polymerization increase as long as new radicals and polymer micelles are formed. Eventually, all surfactant in the system has been absorbed or all initiator molecules have been used up by the polymer particles. At this point in time, the rate of polymerization remains more or less constant. The particle number stabilizes at a rather low value which is only a small fraction, typically about 0.1% of the number of micelles initially present.

The diffusion of the free radicals into the monomer droplets is generally not important to the reaction rate during the growth stage and the rate of monomer diffusion is usually adequate to supply enough monomer to keep the reaction in the particles going.

STAGE III

As the size of the latex particles increases, the size of the monomer droplets decreases and eventually they disappear. At this stage, the reaction mixture consists solely of monomer swollen polymer particles, the so-called latex particles, and dissolved monomers. Since monomer droplets are no longer present, both the monomer concentration and the reaction rate steadily decrease with time. With increasing polymer concentration, some cross-linking reactions and branching can also be expected. The reaction ends when either all monomers are used up or when a second radical diffuses into the polymer particles and causes immediate bimolecular termination by disproportionation or recombination of two radicals. If no termination occurs, the system reaches a conversion of essentially 100 percent. The final (submicroscopic) polymer particles are spherical in shape and have a diameter in the range of 50 – 300 nm which is between the initial micelle and monomer droplet size.

Nitroxide-mediated Radical Polymerization

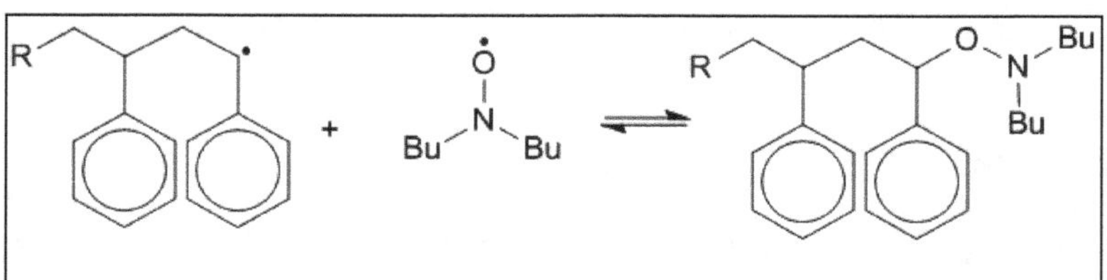

Reversible Termination Reaction Bu = tert-Butyl

Nitroxide-mediated radical polymerization is a method of radical polymerization that makes use of an alkoxyamine initiator to generate polymers with well controlled stereochemistry and a very low polydispersity index. It is a type of reversible-deactivation radical polymerization.

Alkoxyamines

The initiating materials for nitroxide-mediated radical polymerization (NMP) are a family of compounds referred to as alkoxyamines. An alkoxyamine can essentially be viewed as an alcohol bound to a secondary amine by an N-O single bond. The utility of this functional group is that under certain conditions, homolysis of the C-O bond can occur, yielding a stable radical in the form of a 2-center 3-electron N-O system and a carbon radical which serves as an initiator for radical polymerization. For the purposes of NMP, the R groups attached to the nitrogen are always bulky, sterically hindering groups and the R group in the O- position forms a stable radical, generally is benzylic for polymerization to occur successfully. NMP allows for excellent control of chain length and structure, as well as a relative lack of true termination that allows polymerization to continue as long as there is available monomer. Because of this it is said to be "living".

Persistent Radical Effect

The living nature of NMP is due to the persistent radical effect (PRE). The PRE is a phenomenon observable in some radical systems which leads to the highly favored formation of one product to the near exclusion of other radical couplings due to one of the radical species being particularly stable, existing in greater and greater concentrations as the reaction progresses while the other one is transient, reacting quickly with either itself in a termination step or with the persistent radical to form a desired product. As time goes on, a higher concentration of the persistent radical is present, which couples reversibly with itself, meaning that any of the transient radical still present tends to couple with the persistent radical rather than itself due to greater availability. This leads to a greater proportion of cross-coupling than self-coupling in radical species.

In the case of a nitroxide-mediated polymerization reaction, the persistent radical is the nitroxide species and the transient radical is always the carbon radical. This leads to repeated coupling of the nitroxide to the growing end of the polymer chain, which would ordinarily be considered a termination step, but is in this case reversible. Because of the high rate of coupling of the nitroxide to the growing chain end, there is little coupling of two active growing chains, which would be an irreversible terminating step limiting the chain length. The nitroxide binds and unbinds to the growing chain, protecting it from termination steps. This ensures that any available monomer can be easily scavenged by active chains. Because this polymerization process does not naturally self-terminate, this polymerization process is described as "living," as the chains continue to grow under suitable reaction conditions whenever there is reactive monomer to "feed" them. Because of the PRE, it can be assumed that at any given time, almost all of the growing chains are "capped" by a mediating nitroxide, meaning that they dissociate and grow at very similar rates, creating a largely uniform chain length and structure.

Nitroxide Stability

As stated above, nitroxide radicals are effective mediators of well-controlled radical polymerization because they are quite stable, allowing them to act as persistent radicals in a reaction mixture. This stability is a result of their unique structure. In most diagrams, the radical is depicted on the oxygen, but another resonance structure exists which is more helpful in explaining their stability in which the radical is on the nitrogen, which has a double bond to the oxygen. In addition to this resonancestability, nitroxides used in NMRP always contain bulky, sterically hindering groups in the R1 and R2 positions. The significant steric bulk of these substituents entirely prevents radical coupling in the N-centered resonance form while significantly reducing it in the O-centered form. These bulky groups contribute stability, but only if there is no resonance provided by allyl or aromatic groups α to the N. These result in decreased stability of the nitroxide, presumably because they offer less sterically hindered sites for radical coupling to take place. The resulting inactivity of the radical makes hemolytic cleavage of the alkoxyamine quite fast in more sterically hindered species.

Nitroxide Choice

The choice of a specific nitroxide species to use has a large effect on the efficacy of an attempted polymerization. An effective polymerization (fast rate of chain growth, consistent chain length) results from a nitroxide with a fast C-O homolysis and relatively few side reactions. A more polar solvent lends itself better to C-O homolysis, so polar solvents which cannot bind to a labile nitroxide are the most effective for NMP. It is generally agreed that the structural factor that has the greatest effect on the ability of a nitroxide to mediate a radical polymerization is steric bulk. Generally speaking, greater steric bulk on the nitroxide leads to greater strain on the alkoxyamine, leading to the most easily broken bond, the C-O single bond, cleaving homolytically.

Ring Size

In the case of cyclic nitroxides, five-membered ring systems have been shown to cleave more slowly than six-membered rings and acyclic nitroxides with t-butyl moieties as their R groups cleaved fastest of all. This difference in the rate of cleavage was determined to result not from a difference in C-O bond lengths, but in the difference of C-O-N bond angle in the alkoxyamine. The smaller the bond angle the greater the steric interaction between the nitroxide and the alkyl fragment and the more easily the initiator species broke apart.

Steric Bulk

The efficiency of polymerization increases more and more with increased steric bulk of the nitroxide up to a point. TEMPO ((2,2,6,6-Tetramethylpiperidin-1-yl)oxyl) is capable of inducing the polymerization of styrene and styrene derivatives fairly easily, but is not sufficiently labile to induce polymerization of butyl acrylate under most conditions. TEMPO derivatives with even bulkier groups at the positions α to N have a rate of homolysis great enough to induce NMP of butyl acrylate, and the bulkier the α groups, the faster polymerization occurs. This indicates that the steric bulk of the nitroxide fragment can be a good indicator of the strength of an alkoxyamine initiator, at least up to a point. The equilibrium of its homolysis and reformation favors the radical form to the extent that recombination to reform an alkoxyamine over the course of NMP occurs too slowly to maintain control of chain length.

Preparation Methods

Because TEMPO, which is commercially available, is a sufficient nitroxide mediator for the synthesis of polystyrene derivatives, the preparation of alkoxyamine initiators for NMP of copolymers is in many cases a matter of attaching a nitroxide group (TEMPO) to a specifically synthesized alkyl fragment. Several methods have been reported to achieve this transformation.

Jacobsen's Catalyst

Jacobsen's catalyst is a manganese-based catalyst commonly used for the stereoselective epoxidation of alkenes. This epoxidation proceeds by a radical addition mechanism, which can be taken advantage of by introducing the radical TEMPO group into the reaction mixture. After treatment with a mild reducing agent such as sodium borohydride, this yields the product of a Markovnikov addition of nitroxide to the alkene. Jacobsen's catalyst is fairly mild, and a wide variety of functionalities on the alkene substrate can be tolerated. Practical yields are not necessarily as high as those reported by Dao et al., however.

Hydrazine

An alternative method is to react a substrate with a C-Br bond at the desired location of the nitroxide with hydrazine, generating an alkyl substituted hydrazine which is then exposed to a nitroxide radical and a mild oxidating agent such as lead dioxide. This generates a carbon-centered radical which couples with the nitroxide to generate the desired alkoxyamine. This method has the disadvantage of being relatively inefficient for some species, as well as the inherent danger of having to work with extremely toxic hydrazine and the inconvenience of having to run reactions in inert atmosphere.

Treatment of Aldehydes with Hydrogen Peroxide

Yet another published alkoxyamine synthesis involves treatment of aldehydes with hydrogen peroxide, which adds to the carbonyl group. The resulting species rearranges in situ in the presence of CuCl forming formic acid and the desired alkyl radical, which couples with tempo to produce the target alkoxyamine. The reaction appears to give fairly good yields and tolerates a variety of functional groups in the alkyl chain.

Electrophilic Bromination and Nucleophilic Attack

A synthesis has been described by Moon and Kang consisting of a one-electron reduction of a nitroxide radical in metallic sodium to yield a nucleophilic nitroxide. The nitroxide nucleophile is then added to an appropriate alkyl bromide, yielding the alkoxyamine by a simple SN2 reaction. This technique has the advantage of requiring only the appropriate alkyl bromide to be synthesized without requiring inconvenient reaction conditions and extremely hazardous reagents like Braso et al.'s method.

Suspension Polymerization

Suspension polymerization is a heterogeneous radical polymerization process.

Suspension polymerization is used for the commercial manufacture of many important polymers including poly(vinyl chloride), poly(methyl methacrylate), expandable polystyrene, styrene–acrylonitrile copolymers and a variety of ion exchange resins. In suspension polymerization, drops of a monomer-containing phase are dispersed in a continuous liquid phase and polymer is produced inside the drops. In many cases, the monomer contains no diluent and the chemical reactions that occur inside the drops are very similar to those that are found in bulk polymerization. In most suspensions, polymer is formed via a chain reaction mechanism that includes the following steps:

Initiation: $$\begin{aligned} &I \rightarrow 2A^* \\ &A^* + M \rightarrow AM^* \end{aligned}$$

Propagation: $AM_n^* + M \rightarrow AM_{n+1}^*$

Termination: $$\begin{aligned} &AM_n^* + AM_m^* \rightarrow AM_{n+m}A \\ &AM_n^* + AM_m^* \rightarrow AM_n + AM_m \end{aligned}$$

Transfer: $AM_n^* + T \rightarrow AM_n + T^*$

Here, M is the monomer and A* could be an anion, a cation or a free radical. In most industrial processes M is a vinyl compound and a free-radical chain mechanism is used. Then, the growing polymer chains, AM_n^*, are written as $AM_n^{\bullet}$. That species has a short life-time (usually << 1 sec) and completed polymer molecules are formed throughout the process. The generation of radicals, $A^{\bullet}$, is usually induced by thermal decomposition of an organic initiator, I, that is soluble in the monomer. Organic peroxides are often used as initiators. T represents any species that reacts as a chain transfer agent. T can be monomer, polymer, a solvent or a species that is added specifically to function as a chain transfer agent. If T* is sufficiently active it can behave as A* and initiate a new polymer chain. In some cases, such as the polymerization of vinyl chloride, chain transfer to monomer is significant and it has a major effect on the average molecular weight of the polymer.

In most industrial suspension polymerization agitated batch (or semi-batch) reactors are used and the continuous phase is aqueous. That is advantageous because the process is often exothermic and good heat transfer from the reactor is required. The ratio of surface area to volume is relatively high for small drops so that the rate of heat transfer to the aqueous phase is high. Although drop viscosity may increase substantially, the overall viscosity of the suspension is usually much lower than that which is encountered in the equivalent bulk polymerization. Consequently, agitation of the reactor contents is possible and heat transfer via the aqueous phase to the reactor wall is good. Also, high conversions of monomer to polymer can be achieved inside the drops whereas, in bulk polymerization, increasing viscosity of the polymer-monomer solution often limits the extent of monomer conversion.

Bulk copolymerization may become difficult to control if cross-linking or copolymer precipitation occurs; then, a suspension process may then be the only feasible way in which the copolymerization can be carried out.

Suspension polymerization is particularly useful when the final polymer is required to be in the form of small "beads" (which often have the same size distribution as the drops from which they are formed). However, product contamination can be a problem if the drop stabilizers cannot be removed. Suspension polymerization usually requires larger reactor volumes than bulk processes because the vessels are usually half full with water.

The attainment of high monomer conversion can affect the reaction kinetics. From the reaction scheme shown above, it can be shown that the rate of homogeneous polymerization is given by the expression:

$$R_p = k_p C_M \left(\frac{2 f k_d C_1}{k_t} \right)^{1/2}$$

Where *CI* is the concentration of the initiator and *CM* is the monomer concentration. Here, *kp* is the propagation rate coefficient, *kd* is the initiator decomposition rate coefficient and *f* is an efficiency factor. In equation, the overall chain termination rate coefficient, *kt*, is derived from the rate coefficients of the two chain termination steps that are shown in the above reaction scheme. At high polymer concentrations, chain termination is often diffusion-controlled and the value of *kt* diminishes substantially. Radical diffusion can depend on solution viscosity, polymer volume fraction and polymer molecular weight. The latter three entities are interrelated in complicated ways but the effects of viscosity on polymerization rate can be distinguished from the effects of polymer volume fraction. The value of *f* may depend on polymer content and the value of *kp* may also decrease. From equation, it can be seen that the reduction in *kt* leads to an increase in the polymerization rate, a phenomenon often described as a "gel effect".

Suspending Agents

In the absence of a drop stabilizer, the suspension would be unstable and the monomer/polymer drops would coalesce and become large. That is undesirable because, often, it is necessary to obtain a specific size distribution for the final polymer particles. Therefore, control of drop size, and of drop stability, during polymerization becomes important. Drop stability depends largely on the nature of the drop stabilizer (or suspending agent). Adsorption of stabilizer molecules on the outer drop surface can reduce the interfacial tension and, hence, lower the energy required for drop formation. However, drop stability against coalescence depends largely on the ability of the stabilizer to form a protective film at the interface. Increasing the stabilizer concentration continues to improve the elastic properties of the drops until a "critical surface coverage" is attained; further increases then have a very little effect on the drop stability.

Water-miscible polymers, both naturally-occurring and synthetic, are often used as drop stabilizers. Initially, these materials are dispersed in the continuous phase; subsequent migration to the surfaces of newly-created monomer drops may be rapid but the development of drop stability may be slow because rearrangement of stabilizer molecules on the drop surface is necessary . When partially hydrolyzed polyvinyl acetate (PVA) is used as a stabilizer its behavior depends on the extent to which the acetate groups are hydrolyzed. Good drop stabilization can be achieved in aqueous media when the degree of hydrolysis (DH) is between 70% and 80 %; then, drops can retain their integrity even when agitation levels are reduced. PVAs with a DH less than 60% are poor drop stabilizers in aqueous media but they can affect polymer morphology inside the (non-aqueous) drops. That is important in the suspension polymerization of vinyl chloride (VC). In that case, small particles of poly(vinyl chloride) (PVC) precipitate inside the monomer drops because PCV and VC are almost immiscible. Therefore, a mixture of two stabilizers is often used; a "primary stabilizer" which protects the drops from coalescence and a "secondary stabilizer" which effects the behavior of the PVC particles inside the drops and increases polymer porosity. The addition of a secondary stabilizer can also effect the particle size distribution of the polymer particles. PVC porosity can also be increased by using non-ionic surfactants as secondary stabilizers. PVA can become grafted onto polymer that is formed inside the drops so that a "skin" forms on the final particle surface . Formation of that skin, which is difficult to remove, can affect the final polymer properties. With some monomers, product contamination can be avoided by using alternative suspending agents such as salts of polymethacrylic acid which are not grafted on the particle surface and can be removed from the final polymer product with an aqueous wash. If the initiator in suspension polymerization is slightly soluble in water then simultaneous emulsion polymerization may occur when free stabilizer remains in the continuous phase.

Suspension co-polymerization

Functional groups can be introduced via co-polymerization with appropriate monomers but control and prediction of co-polymer compositions in suspension polymerization can be difficult if one, or more, of the monomers is partially soluble in the continuous phase. Then, the actual monomer concentrations in the drops may be unknown so that idealized relationships for predicting co-polymer compositions, which apply to homogeneous systems, are of little use unless appropriate partition coefficients for the two phases are available. Apparent reactivity ratios, obtained directly from suspension polymerization experiments will be different from those expected for the equivalent bulk processes if some monomer migrates to the continuous phase. In some cases, when the continuous phase is aqueous, models that allow for water solubility of monomers have been developed.

Drop Formation and Stability

Control of drop size distribution in suspension polymerisation can be important. In many cases, the average drop diameters (and final average particles sizes) lie between 10 and 100 μm but larger diameters might be produced if the polymer particles are to be used directly as beads.

The physical conditions in a suspension polymerization reactor affect the drop size distribution significantly. Drop breakage in agitated suspensions can be caused either by frictional forces (via viscous shear) or by inertial forces (via turbulence) . In industrial suspension polymerisation, the volume fraction of dispersed phase is usually high and drop break-up is accompanied by drop coalescence. Thus, the average drop size and the drop size distribution are both influenced by drop breakage and by drop coalescence.

Suspensions with Turbulent Flow

In large vessels, agitated aqueous suspensions are often turbulent. If turbulence is isotropic and the diameter exceeds the Kolmogorov length then turbulent pressure fluctuations will cause drop breakage. By applying some simplifying assumptions, the average drop size is sometimes given by equation.

$$\frac{d_{32}}{D} = a(1+b\phi)We^{-0.6}$$

Here, d_{32} is the Sauter–mean drop diameter, φ is the volume fraction of the dispersed phase, D is the impeller diameter and a and b are constants . The Weber number, We, is given by,

$$We = \frac{\rho_m N^2 D^3}{\sigma}$$

Where N is the stirrer speed, ρ_m is the dispersion density and σ is the interfacial tension.

Equation reflects a balance between inertial forces and interfacial forces but, even when drops are formed via turbulence, application of equation is limited to suspensions in which the dispersed phase has a low viscosity and the drop concentration is low (so that drop coalescence is not significant). Wang and Calabrese. showed that, even when turbulence is important, drop break-up can be opposed by both interfacial forces and viscous forces and that the influence of interfacial tension on drop breakage decreases as the dispersed-phase viscosity increases. Drop sizes can take some time to be established and, if polymerisation occurs during that time, the drop viscosity may increase, consequently rates of drop break-up and coalescence will be reduced. Therefore, drop breakage can be a complex process and the polymer particles can have a broad size distribution. In the suspension polymerisation of methyl methacrylate drop viscosity increases significantly and four separate stages have been identified in the drop formation process. In the suspension polymerization of styrene, Konno et al. found that drop coalescence was important and that the Sauter mean diameter increased as the polymer viscosity increased. They also concluded that the stabilizer does not effectively prevent the coalescence of drops with diameters that are larger than those predicted from Weber Number correlations.

The drop coalescence rate can be related to the drop collision frequency and to the coalescence efficiency. Coalescence may occur if drops adhere for sufficient time to allow them to deform, and to permit drainage of the continuous phase that is trapped between them . By taking account of these events, expressions can be obtained for the coalescence efficiency.

Copolymerization

A homopolymer is defined as a polymer which is obtained by linking together only similar types of molecules (or monomers), whereas a copolymer is obtained by linking two or more different types of monomers in the same chain. There are many techniques of polymerization such as condensation, polymerization and copolymerization.

When a mixture of more than one (or different) monomeric species is allowed to polymerize and form a copolymer , then we call this process as copolymerization. Example- Nylon 66 is a copolymer of hexamethylenediamine and adipic acid. There are four different types of copolymers existing today:

- Alternating copolymers: The copolymers which have two units placed at alternating positions. For example, we place a unit X then Y, then X and then again Y and the chain goes on.

—— A —— B —— A —— B ——

The above figure shows the arrangement of alternating copolymers.

- Periodic copolymers: The units of both X and Y is arranged in a repetitive sequence. The arrangement is as shown below:

$$\left(A - B - A - B - B - A - A - A \right)_n$$

- Statistical copolymers: The arrangement of polymers in this case follows a statistical rule. When the probability of finding a given monomer at a particular point is equal to the mole fraction of that monomer in the same chain, then the polymer is referred to as a random copolymer. It has the following arrangement:

$$- A - B - B - B - A - B - A - B -$$

- Block copolymers: When two or more polymers are linked by homopolymers then they are called as block copolymers.

$$- B - B - B - B - A - A - A - A -$$

The above arrangement shows the structure of the block copolymers.

The Copolymer Equation

It can be shown that the rate of change of monomer concentration in any copolymerization is given by the equation:

$$\frac{d[M_1]}{d[M_2]} = \frac{[M_1]}{[M_2]} \cdot \frac{r_1[M_1]+[M_2]}{r_2[M_2]+[M_1]}$$

where $[M_1]$ and $[M_2]$ are the concentrations of monomers 1 and 2 at any instant and r_1 and r_2., are reactivity ratios. The reactivity ratios represent the rate at which one type of growing chain end adds on to a monomer of the same structure relative to the rate at which it adds on to the alternative monomer. The copolymer equation can be used to predict chain structure in the three different ways, already mentioned.

The formation of regular alternating copolymers of the type ABAB is favoured when each growing radical prefers to add to monomer of the opposite type. In this case,

$$r_1 \approx r_2 \approx 0$$

and Equation $\frac{d[M_1]}{d[M_2]} = \frac{[M_1]}{[M_2]} \cdot \frac{r_1[M_1]+[M_2]}{r_2[M_2]+[M_1]}$ therefore becomes

$$\frac{d[M_1]}{d[M_2]} = 1 \text{ or } d[M_1] = d[M_2]$$

In other words both monomers will disappear from the reaction vessel at the same rate.

An ideal copolymer will tend to form when each type of chain end shows an equal preference for adding on to either monomer. In this case,

$$r_1 = \frac{1}{r_2}$$

and the copolymer equation becomes

$$\frac{d[M_1]}{d[M_2]} = r_1 . \frac{[M_1]}{[M_2]}$$

Hence composition depends on the relative amounts of monomer present at any time and the relative reactivities of the two monomers.

Finally, block copolymers are formed when the growing chain end has a marked preference for adding on to the same kind of monomer. In this case,

$$r_1 > 1$$

$$r_2 > 2$$

As can be seen from Table, this is rarely achieved in free radical copolymerization. However, it is possible to form block structures in anionic polymerization simply by feeding different monomers to the living polymer. Step growth copolymerizations produce ideal (random) copolymers since in this special case $r_1 = r_2 = 1$.

Table, Reactivity ratios for free radical chain growth polymerization:

Monomer 1	Monomer 2	r_1	r_2
acrylonitrile	1,3-butadiene	0.02	0.3
	methyl methacrylate	0.15	1.22
	styrene	0.04	0.40
	vinyl acetate	4.2	0.05
	vinyl chloride	2.7	0.04
1,3-butadiene	methyl methacrylate	0.75	0.25
	styrene	1.35	0.78
	vinyl chloride	8.8	0.035
methyl methacrylate	styrene	0.46	0.52
	vinyl acetate	20	0.015
	vinyl chloride	10	0.1
styrene	vinyl acetate	55	0.01
	vinyl chloride	17	0.02
vinyl acetate	vinyl chloride	0.23	1.68

Commercial Copolymers

The main reason for copolymerizing different monomers is to adjust the physical properties of a given homopolymer to meet a specific demand. SBR elastomer, for example Table, based on 24 wt% styrene monomer shows better mechanical properties and better resistance to degradation than polybutadiene alone. By increasing the styrene content to 35 per cent, a high hysteresis (energy absorbing) material ideal for tyre treads is produced. Another example is nitrile rubber, which is produced by a free radical emulsion copolymerization of butadiene and acrylonitrile to make an oil-resistant rubber suitable for oil and petrol lines.

A second reason for copolymerization is to enhance the chemical reactivity of a polymer, particularly to aid crosslinking. Conventional vulcanization in rubbers is brought about by forming sulphur crosslinks at or near double bonds in the chain. In polyisobutylene where the main chain repeat unit is there are no such bonds. So the isobutylene monomer is copolymerized with a few weight percent isoprene units to make IIR (butyl rubber) which can be vulcanized easily.

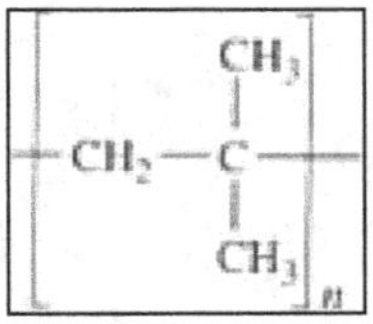

This is also the reason why EPDM rubber consists of no less than three different monomer units copolymerized together (ethylene, propylene and a diene) using Ziegler-Natta catalysts. The copolymer structure is random, so crystallinity is low and the material behaves like a rubber when vulcanized across the diene double bonds.

To show the dramatic effect of copolymer structure on physical properties, consider the change from random SBR copolymer to a block copolymer of exactly the same chemical composition but where the styrene and butadiene parts are effectively homopolymer chains linked at two points:

$$[S]_{n1} - [S]_{n2} - [S]_{n}$$

The material behaves like a vulcanized butadiene rubber without the need for chemical crosslinking since the styrene chains segregate together to form small islands or domains within the structure. Such so-called thermoplastic elastomers (TPEs) today form an important growth area for new polymers because of the process savings in manufacture that can be achieved with their use.

Among rigid thermoplastics, the most widely used copolymers are those of styrene and they include ABS, HIPS and SAN. Both HIPS and ABS are graft block copolymers where the elastomeric side chains are deliberately introduced to improve the toughness of the material.

Step-growth Polymerization

A step-growth polymerization is a stepwise reaction between bi-functional or multi-functional monomers in which a high-molecular-weight polymer is formed after a large number of steps. Many naturally and synthetic polymers are produced by step-growth polymerization including polyesters, polyethers, urethanes, epoxies, and polyamides (see table below). Two well-known examples are the reaction of dicarboxylic acids with diamines to form polyamides (Nylon) and the reaction of organic diacids with alcohols to form polyesters, like polyethylene terephthalate (PET). Due to the nature of the polymerization mechanism, the reaction has to proceed for a long time to achieve high molecular weight polymers. The easiest way to visualize the step-growth mechanism is a crowd of people reaching out to hold their hands to form human chains — each person has two hands (= two reactive sites).

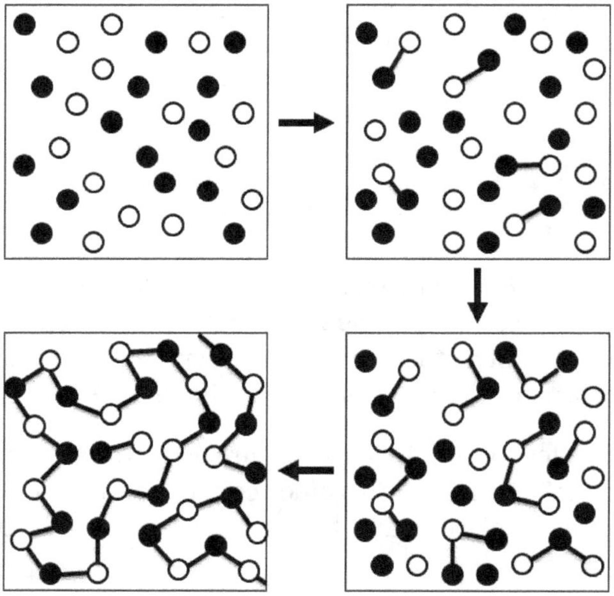

If the monomers have more than two reactive sites branched polymers are produced. Some important step-growth polymerizations are shown below.

EXAMPLES OF STEP-GROWTH POLYMERIZATIONS		
Functional Groups	Linkage	Polymer Type
-OH + -COOH	-C(=O)-O-	Polyester
-OH + -NCO	-O-C(=O)-NH-	Polyurethane
$-NH_2$ + -NCO	-NH-C(=O)-NH-	Polyurea
$-NH_2$ + -COOH	-NH-C(=O)-	Polyamide
-OH + -OH	-O-	Polyether

The two most important step-growth polymerizations are *condensation* and *addition* polymerization. In the case of a condensation reaction, two monomers combine with the loss of a small molecule, usually an alcohol, a water molecule or an acid, whereas an addition reaction involves only the rearrangement of the electrons of a double bond to form a single bond with another molecule.

Classes of Step-growth Polymers

Classes of step-growth polymers are:

- Polyester has high glass transition temperature T_g and high melting point T_m, good mechanical properties to about 175 °C, good resistance to solvent and chemicals. It can exist as fibers and films. The former is used in garments, felts, tire cords, etc. The latter appears in magnetic recording tape and high grade films.
- Polyamide (nylon) has good balance of properties: high strength, good elasticity and abrasion resistance, good toughness, favorable solvent resistance. The applications of polyamide include: rope, belting, fiber cloths, thread, substitute for metal in bearings, jackets on electrical wire.
- Polyurethane can exist as elastomers with good abrasion resistance, hardness, good resistance to grease and good elasticity, as fibers with excellent rebound, as coatings with good resistance to solvent attack and abrasion and as foams with good strength, good rebound and high impact strength.
- Polyurea shows high T_g, fair resistance to greases, oils, and solvents. It can be used in truck bed liners, bridge coating, caulk and decorative designs.
- Polysiloxane are available in a wide range of physical states—from liquids to greases, waxes, resins, and rubbers. Uses of this material are as antifoam and release agents, gaskets, seals, cable and wire insulation, hot liquids and gas conduits, etc.
- Polycarbonates are transparent, self-extinguishing materials. They possess properties like crystalline thermoplasticity, high impact strength, good thermal and oxidative stability. They can be used in machinery, auto-industry, and medical applications. For example, the cockpit canopy of F-22 Raptor is made of high optical quality polycarbonate.
- Polysulfides have outstanding oil and solvent resistance, good gas impermeability, good resistance to aging and ozone. However, it smells bad, and it shows low tensile strength as well as poor heat resistance. It can be used in gasoline hoses, gaskets and places that require solvent resistance and gas resistance.
- Polyether shows good thermoplastic behavior, water solubility, generally good mechanical properties, moderate strength and stiffness. It is applied in sizing for cotton and synthetic fibers, stabilizers for adhesives, binders, and film formers in pharmaceuticals.

- Phenol formaldehyde resin (Bakelite) have good heat resistance, dimensional stability as well as good resistance to most solvents. It also shows good dielectric properties. This material is typically used in molding applications, electrical, radio, televisions and automotive parts where their good dielectric properties are of use. Some other uses include: impregnating paper, varnishes, decorative laminates for wall coverings.
- Poly-Triazole polymers are produced from monomers which bear both an alkyne and azide functional group. The monomer units are linked to each other by the a 1,2,3-triazole group; which is produced by the 1,3-Dipolar cycloaddition, also called the Azide-alkyne Huisgen cycloaddition. These polymers can take on the form of a strong resin, or a gel. With oligopeptide monomers containing a terminal alkyne and terminal azide the resulting clicked peptide polymer will be biodegradable due to action of endopeptidases on the oligopeptide unit.

Branched Polymers

A monomer with functionality of 3 or more will introduce branching in a polymer and will ultimately form a cross-linked macrostructure or network even at low fractional conversion. The point at which a tree-like topology transits to a network is known as the gel point because it is signalled by an abrupt change in viscosity. One of the earliest so-called thermosets is known as bakelite. It is not always water that is released in step-growth polymerization: in acyclic diene metathesis or ADMET dienes polymerize with loss of ethylene.

Kinetics

The kinetics and rates of step-growth polymerization can be described using a polyesterification mechanism. The simple esterification is an acid-catalyzed process in which protonation of the acid is followed by interaction with the alcohol to produce an ester and water. However, there are a few assumptions needed with this kinetic model. The first assumption is water (or any other condensation product) is efficiently removed. Secondly, the functional group reactivity's are independent of chain length. Finally, it is assumed that each step only involves one alcohol and one acid.

$$\frac{1}{1-p^{n-1}} = 1 + (n-1)kt[\mathrm{COOH}]^{n-1}$$

This is a general rate law degree of polymerization for polyesterification where n= reaction order.

Self-catalyzed Polyesterification

If no acid catalyst is added, the reaction will still proceed because the acid can act as its own catalyst. The rate of condensation at any time t can then be derived from the rate of disappearance of -COOH groups and

$$rate = \frac{-d[\text{COOH}]}{dt} = k[\text{COOH}]^2[\text{OH}]$$

The second-order [\ce{COOH}] term arises from its use as a catalyst, and k is the rate constant. For a system with equivalent quantities of acid and glycol, the functional group concentration can be written simply as:

$$rate = \frac{-d[\text{COOH}]}{dt} = k[\text{COOH}]^3$$

After integration and substitution from Carothers equation, the final form is the following,

$$\frac{1}{(1-p)^2} = 2kt[\text{COOH}]^2 + 1 = X_n^2$$

For a self-catalyzed system, the number average degree of polymerization (Xn) grows proportionally with $\sqrt{t}$.

External Catalyzed Polyesterification

The uncatalyzed reaction is rather slow, and a high X_n is not readily attained. In the presence of a catalyst, there is an acceleration of the rate, and the kinetic expression is altered to,

$$\frac{-d[\text{COOH}]}{dt} = k[\text{COOH}][\text{OH}]$$

Which is kinetically first order in each functional group. Hence,

$$\frac{-d[\text{COOH}]}{dt} = k[\text{COOH}]^2$$

and integration gives finally

$$\frac{1}{1-p} = 1 + [\text{COOH}]kt = X_n$$

For an externally catalyzed system, the number average degree of polymerization grows proportionally with t.

Molecular Weight Distribution in Linear Polymerization

The product of a polymerization is a mixture of polymer molecules of different molecular weights. For theoretical and practical reasons it is of interest to discuss the distribution of molecular weights in a polymerization. The molecular weight distribution (MWD) had been derived by Flory by a statistical approach based on the concept of equal reactivity of functional groups.

Probability

Step-growth polymerization is a random process so we can use statistics to calculate the probability of finding a chain with x-structural units ("x-mer") as a function of time or conversion.

$$x\,\text{AA} + x\,\text{BB} \rightarrow \text{AA-(BB-AA)}_{x-1}\text{-BB}$$

$$x\,\text{AB} \rightarrow \text{A-(B-A)}_{x-1} - \text{B}$$

Probability that an 'A' functional group has reacted:

$$p^{x-1}$$

Probability of finding an 'A' unreacted:

$$(1-p)$$

Combining the above two equations leads to:

$$P_x = (1-p)p^{x-1}$$

Where P_x is the probability of finding a chain that is x-units long and has an unreacted 'A'. As x increases the probability decreases.

Number Fraction Distribution

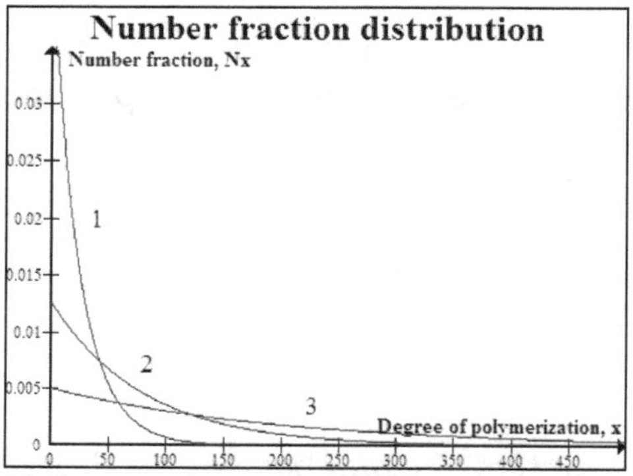

Number-fraction distribution curve for linear polymerization. Plot 1, p=0.9600; plot 2, p=0.9875; plot 3, p=0.9950.

The number fraction distribution is the fraction of x-mers in any system and equals the probability of finding it in solution.

$$\frac{N_x}{N} = (1-p)p^{x-1}$$

Where N is the total number of polymer molecules present in the reaction.

Weight Fraction Distribution

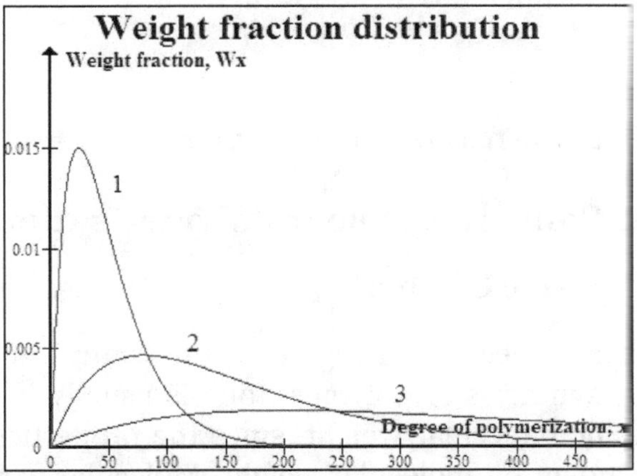

Weight fraction distribution plot for linear polymerization. Plot 1, p=0.9600; plot 2, p=0.9875; plot 3, p=0.9950.

The weight fraction distribution is the fraction of x-mers in a system and the probability of finding them in terms of mass fraction.

$$\frac{W_x}{W_o} = \frac{xN_xM_o}{N_oM_o} = \frac{xN_x}{N_o} = x\frac{N_x}{N}\frac{N}{N_o}$$

Notes:

- M_o is the molar mass of the repeat unit,
- N_o is the initial number of monomer molecules,
- and N is the number of unreacted functional groups.

Substituting from the Carothers equation:

$$X_n = \frac{1}{1-p} = \frac{N_o}{N}$$

We can now obtain:

$$\frac{W_x}{W_o} = x(1-p)^2 p^{x-1}$$

PDI

The polydispersity index (PDI), is a measure of the distribution of molecular mass in a given polymer sample.

$$PDI = \frac{M_w}{M_n}$$

However, for step-growth polymerization the Carothers equation can be used to substitute and rearrange this formula into the following.

$$PDI = 1 + p$$

Therefore, in step-growth when p=1, then the PDI=2.

Molecular Weight Control in Linear Polymerization

Need for Stoichiometric Control

There are two important aspects with regard to the control of molecular weight in polymerization. In the synthesis of polymers, one is usually interested in obtaining a product of very specific molecular weight, since the properties of the polymer will usually be highly dependent on molecular weight. Molecular weights higher or lower than the desired weight are equally undesirable. Since the degree of polymerization is a function of reaction time, the desired molecular weight can be obtained by quenching the reaction at the appropriate time. However, the polymer obtained in this manner is unstable in that it leads to changes in molecular weight because the ends of the polymer molecule contain functional groups that can react further with each other.

This situation is avoided by adjusting the concentrations of the two monomers so that they are slightly nonstoichiometric. One of the reactants is present in slight excess. The polymerization then proceeds to a point at which one reactant is completely used up and all the chain ends possess the same functional group of the group that is in excess. Further polymerization is not possible, and the polymer is stable to subsequent molecular weight changes.

Another method of achieving the desired molecular weight is by addition of a small amount of monofunctional monomer, a monomer with only one functional group. The monofunctional monomer, often referred to as a chain stopper, controls and limits the polymerization of bifunctional monomers because the growing polymer yields chain ends devoid of functional groups and therefore incapable of further reaction.

Quantitative Aspects

To properly control the polymer molecular weight, the stoichiometric imbalance of the bifunctional monomer or the monofunctional monomer must be precisely adjusted. If the nonstoichiometric imbalance is too large, the polymer molecular weight will be too low. It is important to understand the quantitative effect of the stoichiometric imbalance of reactants on the molecular weight. Also, this is necessary in order to know the quantitative effect of any reactive impurities that may be present in the reaction mixture either initially or that are formed by undesirable side reactions. Impurities with A or B functional groups may drastically lower the polymer molecular weight unless their presence is quantitatively taken into account.

More usefully, a precisely controlled stoichiometric imbalance of the reactants in the mixture can provide the desired result. For example, an excess of diamine over an acid chloride would eventually produce a polyamide with two amine end groups incapable of further growth when the acid chloride was totally consumed. This can be expressed in an extension of the Carothers equation as,

$$X_n = \frac{(1+r)}{(1+r-2rp)}$$

where r is the ratio of the number of molecules of the reactants.

$r = \frac{N_{AA}}{N_{BB}}$ were N_{BB} is the molecule in excess.

The equation above can also be used for a monofunctional additive which is the following,

$$r = \frac{N_{AA}}{(N_{BB}+2N_B)}$$

where N_B is the number of monofunction molecules added. The coefficient of 2 in front of N_B is require since one B molecule has the same quantitative effect as one excess B-B molecule.

Multi-chain Polymerization

A monomer with functionality 3 has 3 functional groups which participate in the polymerization. This will introduce branching in a polymer and may ultimately form a cross-linked macrostructure. The point at which this three-dimensional 3D network is formed is known as the gel point, signaled by an abrupt change in viscosity.

A more general functionality factor f_{av} is defined for multi-chain polymerization, as the average number of functional groups present per monomer unit. For a system containing N_o molecules initially and equivalent numbers of two function groups A and B, the total number of functional groups is $N_o f_{av}$.

$$f_{av} = \frac{\sum N_i \cdot f_i}{\sum N_i}$$

And the modified Carothers equation is,

$$x_n = \frac{2}{2-pf_{av}}, \text{ where p equals to } \frac{2(N_o - N)}{N_o \cdot f_{av}}$$

Advances in Step-growth Polymers

The driving force in designing new polymers is the prospect of replacing other materials of construction, especially metals, by using lightweight and heat-resistant polymers.

The advantages of lightweight polymers include: high strength, solvent and chemical resistance, contributing to a variety of potential uses, such as electrical and engine parts on automotive and aircraft components, coatings on cookware, coating and circuit boards for electronic and microelectronic devices, etc. Polymer chains based on aromatic rings are desirable due to high bond strengths and rigid polymer chains. High molecular weight and crosslinking are desirable for the same reason. Strong dipole-dipole, hydrogen bond interactions and crystallinity also improve heat resistance. To obtain desired mechanical strength, sufficiently high molecular weights are necessary, however, decreased solubility is a problem. One approach to solve this problem is to introduce of some flexibilizing linkages, such as isopropylidene, C=O, and SO_2 into the rigid polymer chain by using an appropriate monomer or comonomer. Another approach involves the synthesis of reactive telechelic oligomers containing functional end groups capable of reacting with each other, polymerization of the oligomer gives higher molecular weight, referred to as chain extension.

Aromatic Polyether

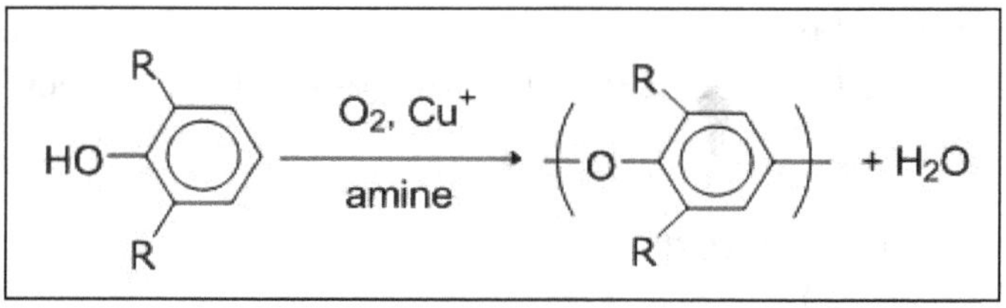

The oxidative coupling polymerization of many 2,6-disubstituted phenols using a catalytic complex of a cuprous salt and amine form aromatic polyethers, commercially referred to as poly(p-phenylene oxide) or PPO. Neat PPO has little commercial uses due to its high melt viscosity. Its available products are blends of PPO with high-impact polystyrene (HIPS).

Polyethersulfone

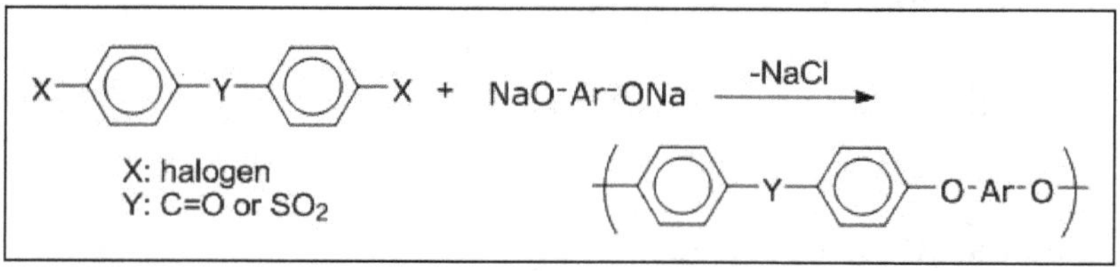

Polyethersulfone (PES) is also referred to as polyetherketone, polysulfone. It is synthesized by nucleophilic aromatic substitution between aromatic dihalides and bisphenolate salts. Polyethersulfones are partially crystalline, highly resistant to a wide range of aqueous and organic environment. They are rated for continuous service at temperatures of 240-280 °C. The polyketones are finding applications in areas like automotive, aerospace, electrical-electronic cable insulation.

Aromatic Polysulfides

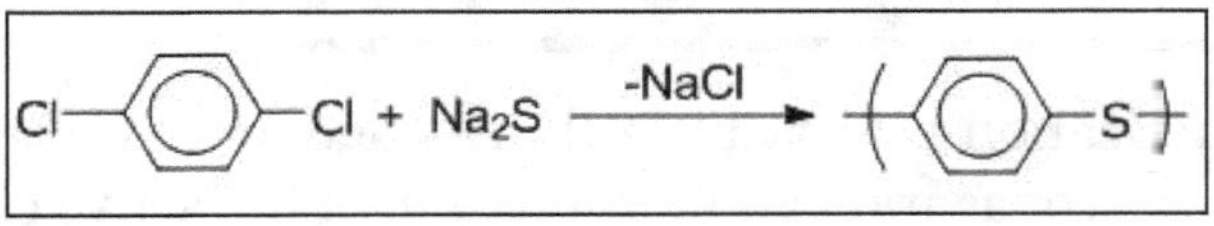

Poly(p-phenylene sulfide) (PPS) is synthesized by the reaction of sodium sulfide with p-dichlorobenzene in a polar solvent such as 1-methyl-2-pyrrolidinone (NMP). It is inherently flame-resistant and stable toward organic and aqueous conditions; however, it is somewhat susceptible to oxidants. Applications of PPS include automotive, microwave oven component, coating for cookware when blend with fluorocarbon polymers and protective coatings for valves, pipes, electromotive cells, etc.

Aromatic Polyimide

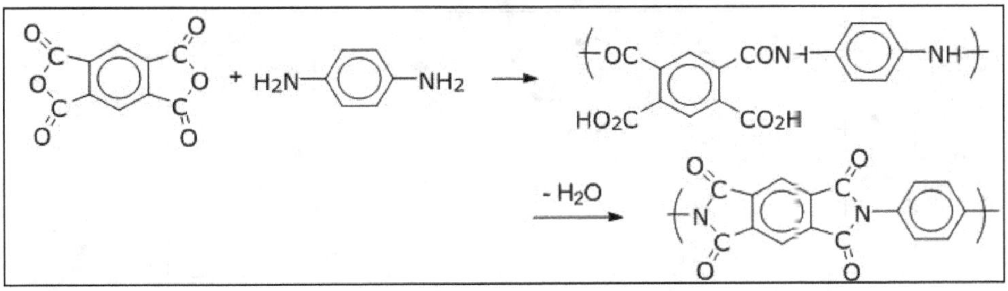

Aromatic polyimides are synthesized by the reaction of dianhydrides with diamines, for example, pyromellitic anhydride with *p*-phenylenediamine. It can also be accomplished using diisocyanates in place of diamines. Solubility considerations sometimes suggest use of the half acid-half ester of the dianhydride, instead of the dianhydride itself. Polymerization is accomplished by a two-stage process due to the insolubility of polyimides. The first stage forms a soluble and fusible high-molecular-weight poly(amic acid) in a polar aprotic solvent such as NMP or N,N-dimethylacetamide. The poly(amic aicd) can then be processed into the desired physical form of the final plymer product (e.g., film, fiber, laminate, coating) which is insoluble and infusible.

Telechelic Oligomer Approach

Telechelic oligomer approach applies the usual polymerization manner except that one includes a monofunctional reactant to stop reaction at the oligomer stage, generally in the 50-3000 molecular weight. The monofunctional reactant not only limits polymerization but end-caps the oligomer with functional groups capable of subsequent reaction to achieve curing of the oligomer. Functional groups like alkyne, norbornene, maleimide, nitrite, and cyanate have been used for this purpose. Maleimide and norbornene end-capped oligomers can be cured by heating. Alkyne, nitrile, and cyanate end-capped oligomers can undergo cyclotrimerization yielding aromatic structures.

Chain-growth Polymerization

Chain growth polymerization is basically a three-stage process, involving initiation of active molecules, their propagation and termination of the active chain ends.

Initiation

Initiation is the mechanism which starts the polymerization process. Vinyl monomers are quite easily polymerized by a variety of activating methods. Styrene, for example, can be converted to solid polymer simply by heating, and ultraviolet light can have exactly the same effect. Usually, however, an activating agent is used. This is an unstable chemical which produces active species that attack the monomer. A good example is benzoyl peroxide which splits up when heated:

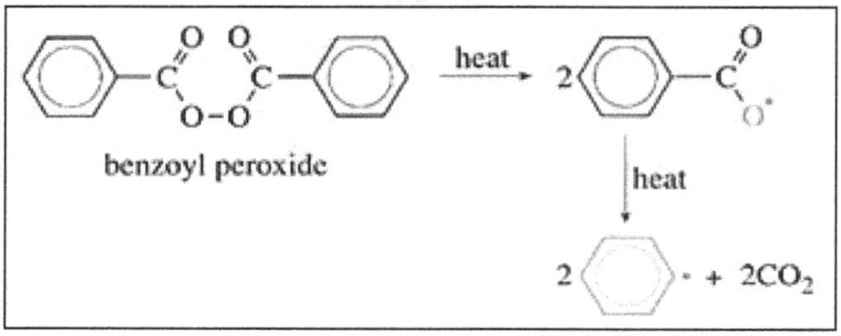

The formulae of the products are written with a dot alongside to show that they are free radicals. A free radical is a molecule in which there is an unpaired electron. This free radical is very reactive and will attack monomer molecules when introduced into a polymerization vessel. Thus, as benzoyl peroxide is added to styrene (a reaction used with GRP), the peroxide splits to make free radicals, which react as follows:

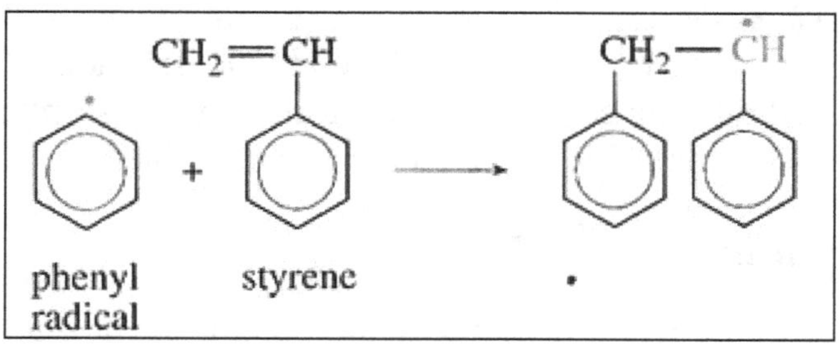

The net result is that the reactants have been linked together but the product is still a radical and so is capable of attacking further monomer molecules. In each instance the attack will lead to a larger molecule but the free radical will be preserved. The reaction is referred to as free radical polymerization.

Free radicals are not the only way of initiating reactions. Charged molecules can often exert the same effect. Ethyllithium, for example, is a relatively unstable molecule which can dissociate to form an ion pair:

$$C_2H_5\,Li \rightarrow C_2H_5^- + Li^+$$

Styrene can also be polymerized by this compound:

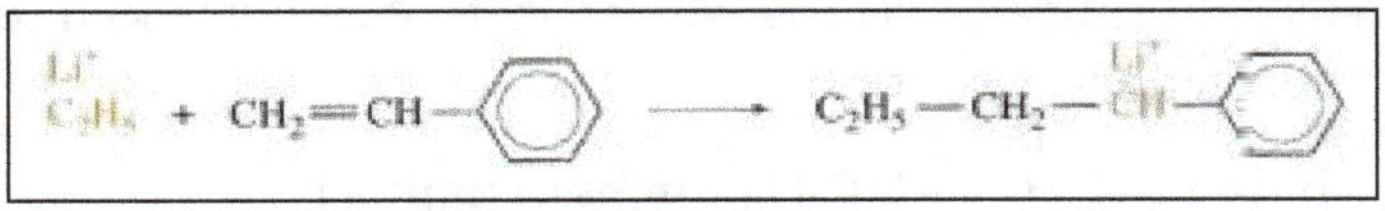

This mechanism is called anionic polymerization.

Propagation

Once a small number of chains have been started, propagation involves successive addition of monomer units to achieve chain growth. At each step the free radical is regenerated as it reacts with the double bond. So in the case of styrene the propagation step is,

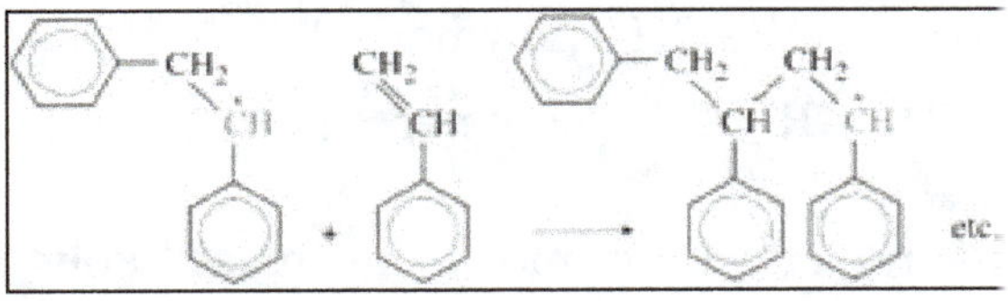

The free radical can also add on in a different way to produce,

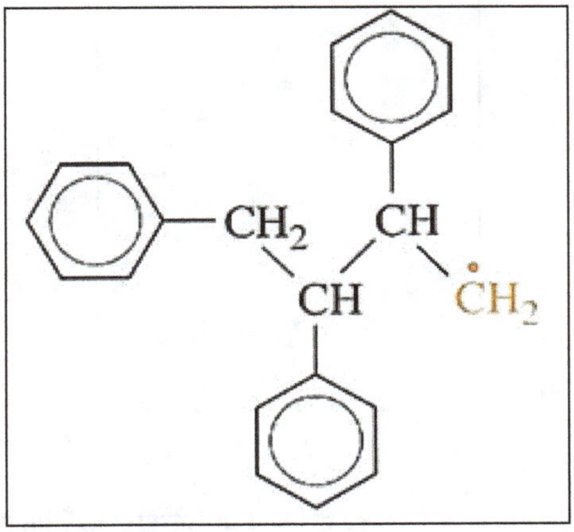

but this process happens only rarely since the free radical is less stable than in the first case.

The junction that is formed normally is known as a head-to-tail link while the abnormal link is head-to-head. The effect is limited to about 1 per cent of the total number of monomer links in normal polystyrene, but it is important because the head-to-head links are weaknesses in the chain. Since they are of higher energy, thermal degradation can start at these defective junctions.

Termination and Transfer

There are basically three ways in which chains terminate.

The first is known as coupling and occurs when two free radicals join together. This can be represented by the general equation

$$(\text{polymer chain 1})-\text{RH}\bullet+\bullet\text{HR}-(\text{polymer chain 2})$$
$$\rightarrow(\text{polymer chain 1})-\text{RH}-\text{RH}-(\text{polymer chain 2})$$

Such a mechanism significantly increases molecular mass, if it results in two polymer chains joining. This is the main mechanism which terminates the polymerization of styrene.

An alternative mechanism that may occur when two radicals interact is known as disproportionation. In this case, one molecule abstracts a hydrogen atom from the other and the other molecule forms a double bond,

$$(\text{polymer 1})-\text{RH}\bullet+\bullet\text{HR}-\text{CH}_2(\text{polymer 2})$$
$$\rightarrow(\text{polymer 1})-\text{RH}_2+\text{R}=\text{CH}-(\text{polymer 2})$$

Disproportionation has no effect on molecular mass. Poly(methyl methacrylate) (PMMA) terminates by a mixture of coupling and disproportionation.

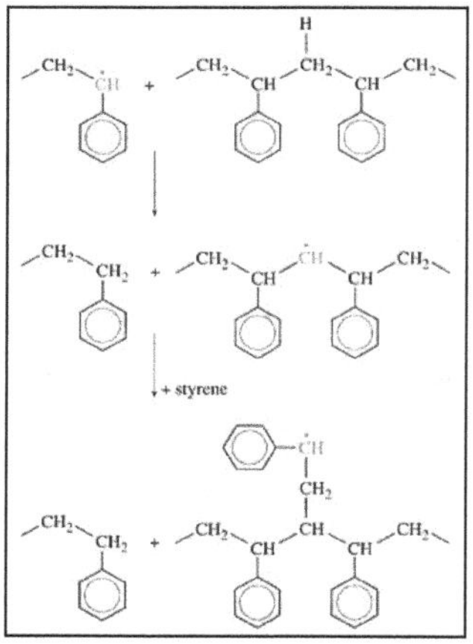

Figure: Formation of branched polymer in free radical polystyrene. The growing chain end abstracts a hydrogen atom from neighbouring polymer (top) and (middle), which then propagates with styrene monomer to form a single branch (bottom).

The third method of termination is chain transfer in which a radical abstracts a hydrogen atom from a neighbouring molecule. In the case of polystyrene the effect will be as shown in Figure, where (a) shows the situation before the interaction and (b) shows the structures after chain transfer in which the radical is transferred to one of the mid-chain carbon atoms. The new radical may now attack further styrene (Figure(c)) but, because it is not on the end of the chain, side branching occurs.

A similar mechanism accounts for the side branches in LDPE where it is a more important mode of termination than in polystyrene. Transfer to monomer; initiator or solvent (if present) can also occur in free radical polymerization, and effectively increases the dispersion of the molecular mass of the final polymer.

If termination is simply by disproportionation, then:

$$n = \frac{K[\mathrm{M}]}{[\mathrm{I}]^{1/2}}$$

where K is a constant, [M] the concentration of monomer, [I] the concentration of initiator, e.g. peroxide, and n the degree of polymerization. The square root arises because two free radicals react together during termination. If termination is by coupling there will be an extra factor of two in the constant compared to disproportionation. So the degree of polymerization or molecular mass can be controlled by varying monomer concentration – for example, by conducting the reaction in solvent – or by varying initiator concentration.

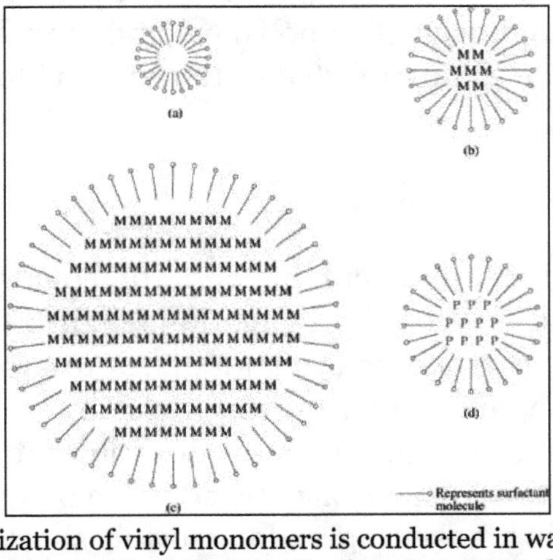

Figure: Emulsion polymerization of vinyl monomers is conducted in water to aid heat dissipation. Soap micelles (a) swell with monomer (b) which migrates from monomer droplets stabilised by soap molecules (c). The process yields a polymer latex (d) which can be used directly or reduced to bulk polymer.

Controlling polymerizations on an industrial scale is of critical importance for molecular mass, and hence the processability and physical properties of the polymer, and one of the most important variables is the temperature of reaction. All polymerizations are exothermic (heat is liberated due to bond formation) and the heat must be conducted away to maintain a uniform reaction temperature. This is much more easily achieved when an inert solvent is used. Another method very commonly used industrially is to emulsify the monomer with a soap and conduct the reaction in water – so-called emulsion polymerization Figure. Since control of molecular mass is so vital, extra aids are used industrially in addition to varying monomer and initiator concentrations. Reactions are 'short stopped' before all monomer is consumed by adding a specific chemical which reacts with free radicals, stopping them dead. Other chemicals can be added to induce transfer reactions, so controlling molecular mass distribution.

Living Polymerization

A Living polymerization is defined as a chain polymerization without irreversible transfer and termination. Living Polymerizations will lead to well-defined polymers only if the following additional prerequisites are fulfilled:

- initiation is fast in comparison with propagation,
- exchange between species of different reactivities is fast in comparison with propagation,
- rate of depropagation is low in comparison with propagation,
- system is sufficiently homogeneous, in sense of the availability of active centers and mixing.

If these specifications are not met, living polymerizations will produce polymers with broader polydispersities and degrees of polymerization much higher than the Δ [M]/[I]o ratio. The proportion of chains affected by transfer and termination increases with the chain length.

Living Polymerization Process

There are four general processes occurring simultaneously during the polymerization process: chain initiation, chain propagation, chain transfer, and chain termination. In simple terms, each polymer chain starts to grow, propagates, and terminates at a certain time, and synchronizing these processes results in chains of similar length, or molecular weight, which is desirable. If the chain initiation rate is slower or comparable to propagation rate, some chains are being initiated while others are rapidly growing, resulting in longer and shorter chains. On the other hand, if initiation is much faster than the propagation, polymer chains start growing simultaneously, and grow uniformly. Now, if the chains are not terminated by any additional mechanism, the only factor defining their growth is the presence of a monomer. Once the monomer is depleted, the growth is complete, with resulting polymer chains of the same length.

Living polymerization, which has been studied for more than 70 years, can follow anionic, cationic, and radical polymerization mechanisms. Popular atom transfer radical polymerization (ATRP) and reversible addition-fragmentation chain transfer (RAFT) are examples of living radical polymerization. Living polymerization allows you to obtain precisely controlled molecular weight and narrow molecular weight distribution, as well as complex polymer architectures.

Techniques

Living Anionic Polymerization

As early as 1936, Karl Ziegler proposed that anionic polymerization of styrene and butadiene by consecutive addition of monomer to an alkyl lithium initiator occurred without chain transfer or termination. Twenty years later, living polymerization was demonstrated by Szwarc through the anionic polymerization of styrene in THF using sodium naphthalenide as celerator.

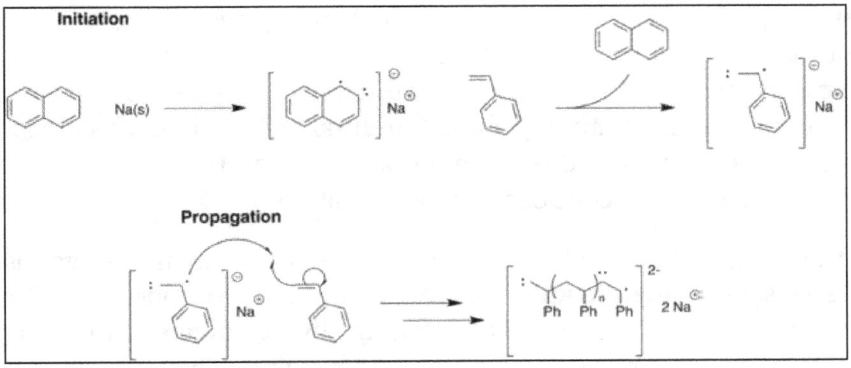

Here, the naphthalene anion acts as the initiator of the polymerization by activating the styrene. However, note that (with no impurities present for quenching and no solvent for chain transfer) there is no route for termination to occur. Therefore, these terminal anions will stay on the ends of the polymer until a quenching agent is introduced.

It is believed that the dianion of the polymer shown above is formed for this reaction, allowing the propagation to occur at either end of the chain. However, notice that there is no termination step (given impurities are not present to quench). This is the basis for anionic living polymerizations, where the terminal radical will exist until free monomer is available for additional propagation, or is quenched from an outside source.

Living α-olefin Polymerization

α-olefins can be polymerized through an anionic coordination polymerization in which the metal center of the catalyst is considered the counter cation for the anionic end of the alkyl chain (through a M-R coordination). Ziegler-Natta initiators were developed in the mid-1950s and are heterogeneous initiators used in the polymerization of alpha-olefins. Not only were these initiators the first to achieve relatively high molecular weight poly(1-alkenes) (currently the most widely produced thermoplastic in the world PE(Polyethylene) and PP (Polypropylene) but the initiators were also capable of stereoselctive polymerizations which is attributed to the chiral Crystal structure of the heterogeneous initiator. Due to the importance of this discovery Ziegler and Natta were presented with the 1963 Nobel Prize in chemistry. Although the active species formed from the Ziegler-Natta initiator generally have long lifetimes (on the scale of hours or longer) the lifetimes of the propagating chains are shortened due to several chain transfer pathways (Beta-Hydride elimination and transfer to the co-initiator) and as a result are not considered living.

Metallocene initiators are considered as a type of Ziegler-Natta initiators due to the use of the two-component system consisting of a transition metal and a group I-III metal co-initiator (for example methylalumoxane (MAO) or other alkyl aluminum compounds). The metallocene initiators form homogeneous single site catalysts that were initially developed to study the impact that the catalyst structure had on the resulting polymers structure/properties; which was difficult for multi-site heterogeneous Ziegler-Natta initiators. Owing to the discrete single site on the metallocene catalyst researchers were able to tune and relate how the ancillary ligand (those not directly involved in the chemical transformations) structure and the symmetry about the chiral metal center affect the microstructure of the polymer. However, due to chain breaking reactions (mainly Beta-Hydride elimination) very few metallocene based polymerizations are known.

By tuning the steric bulk and electronic properties of the ancillary ligands and their substituents a class of initiators known as chelate initiators (or post-metallocene initiators) have been successfully used for stereospecific living polymerizations of alpha-olefins. The chelate initiators have a high potential for living polymerizations because the ancillary ligands can be designed to discourage or inhibit chain termination pathways. Chelate initiators can be further broken down based on the ancillary ligands; ansa-cyclopentyadienyl-amido initiators, alpha-diimine chelates and phenoxy-imine chelates.

- Ansa-cyclopentadienyl-amido (CpA) initiators,

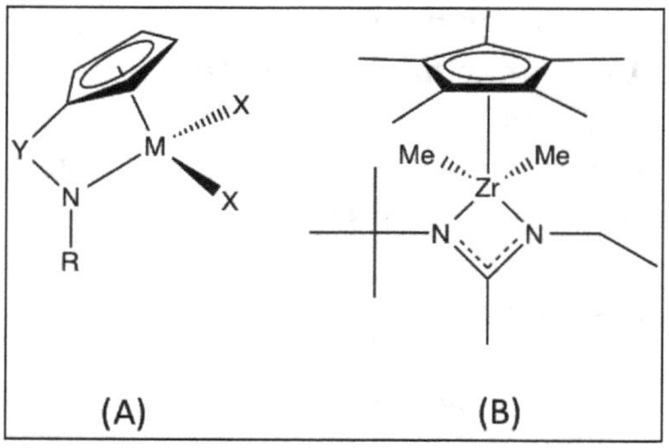

a.) Shows the general form of CpA initiators with one Cp ring and a coordinated Nitrogen b.) Shows the CpA initiator used in the living polymerization of 1-hexene.

CpA initiators have one cyclopentadienyl substituent and one or more nitrogen substituents coordinated to the metal center (generally a Zr or Ti) (Odian). The dimethyl(pentamethylcyclopentyl)zirconium acetamidinate in figure has been used for a stereospecific living polymerization of 1-hexene at −10 deg C. The resulting poly(1-hexene) was isotactic (stereohemistry is the same between adjacent repeat units) confirmed by ^{13}C-NMR. The multiple trials demonstrated a controllable and predictable (from catalyst to monomer ratio) M_n with low Đ. The polymerization was further confirmed to be living by sequentially adding 2 portions of the monomer, the second portion was added after the first portion was already polymerized, and monitoring the Đ and M_n of the chain. The resulting polymer chains complied with the predicted M_n (with the total monomer concentration = portion 1 +2) and showed low Đ suggesting the chains were still active, or living, as the second portion of monomer was added.

- α-diimine chelate initiators,

α-diimine chelate initiators are characterized by having a diimine chelating ancillary ligand structure and which is generally coordinated to a late transition (i.e. Ni and Pd) metal center.

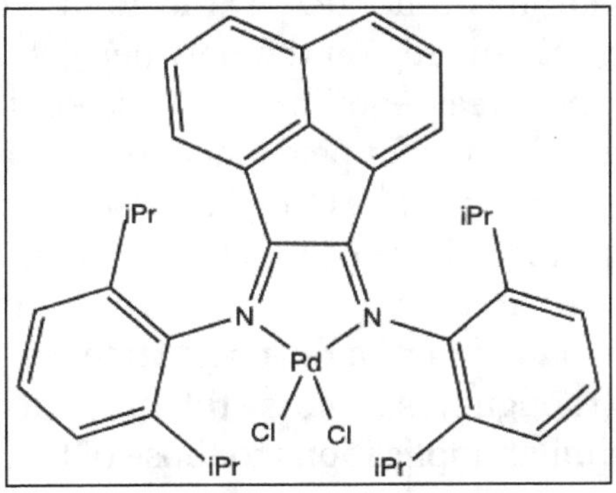

Brookhart et al. did extensive work with this class of catalysts and reported living polymerization for α-olefins and demonstrated living α-olefin carbon monoxide alternating copolymers.

- Phenoxy-imine chelates,

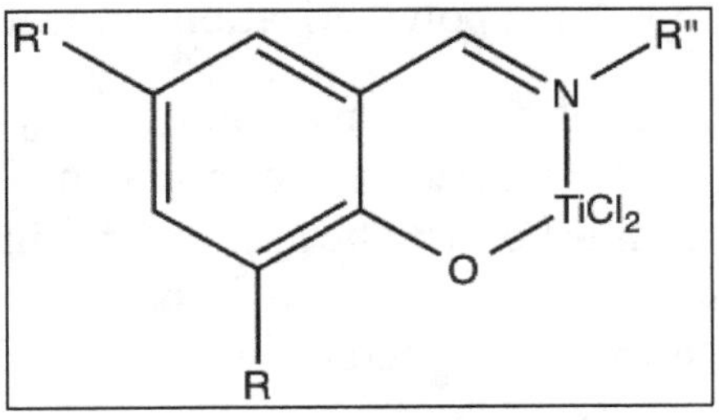

Living Cationic Polymerization

Monomers for living cationic polymerization are electron-rich alkenes such as vinyl ethers, isobutylene, styrene, and N-vinylcarbazole. The initiators are binary systems consisting of an electrophile and a Lewis acid. The method was developed around 1980 with contributions from Higashimura, Sawamoto and Kennedy. Typically, generating a stable carbocation for a prolonged period of time is difficult, due to the possibility for the cation to be quenched by a β-protons attached to another monomer in the backbone, or in a free monomer. Therefore, a different approach is taken.

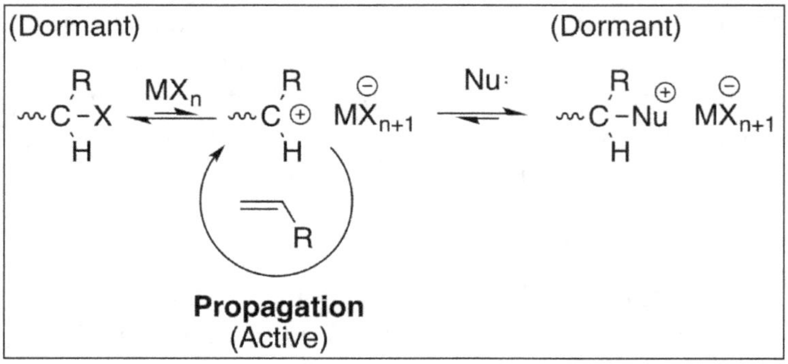

In this example, the carbocation is generated by the addition of a Lewis acid (co-initiator, along with the halogen "X" already on the polymer – see figure), which ultimately generates the carbocation in a weak equilibrium. This equilibrium heavily favors the dormant state, thus leaving little time for permanent quenching or termination by other pathways. In addition, a weak nucleophile (Nu:) can also added to reduce the concentration of active species even further, thus keeping the polymer "living". However, it is important to note that by definition, the polymers described in this example are not technically living due to the introduction of a dormant state, as termination has only been decreased, not eliminated (though this topic is still up for debate). But, they do operate similarly, and are used in similar applications to those of true living polymerizations.

Living Ring-opening Metathesis Polymerization

Given the right reaction conditions ring-opening metathesis polymerization (ROMP) can be rendered living. The first such systems were described by Robert H. Grubbs in 1986 based on norbornene and Tebbe's reagent and in 1978 Grubbs together with Richard R. Schrock describing living polymerization with a tungsten carbene complex.

Generally, ROMP reactions involve the conversion of a cyclic olefin with significant ring-strain (>5 kcal/mol), such as cyclobutene, norbornene, cyclopentene, etc., to a polymer that also contains double bonds. The important thing to note about ring-opening metathesis polymerizations is that the double bond is usually maintained in the backbone, which can allow it to be considered "living" under the right conditions.

For a ROMP reaction to be considered "living", several guidelines must be met:

1. Fast and complete initiation of the monomer. This means that the rate at which an initiating agent activates the monomer for polymerization, must happen very quickly.

2. How many monomers make up each polymer (the degree of polymerization) must be related linearly to the amount of monomer you started with.

3. The dispersity of the polymer must be < 1.5. In other words, the distribution of how long your polymer chains are in your reaction must be very low.

With these guidelines in mind, it allows you to create a polymer that is well controlled both in content (what monomer you use) and properties of the polymer (which can be largely attributed to polymer chain length). It is important to note that living ring-opening polymerizations can be anionic *or* cationic.

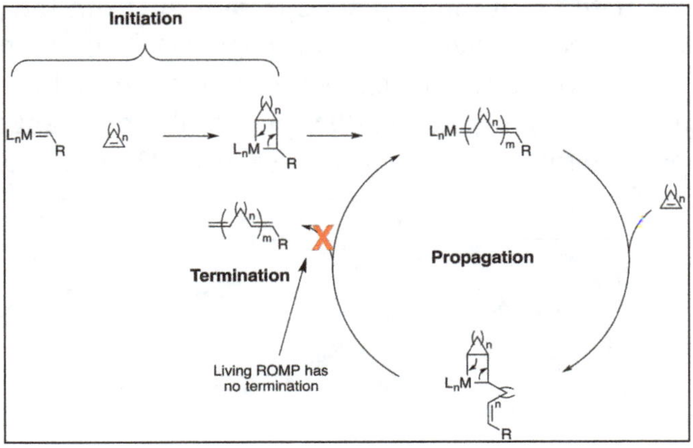

Because living polymers have had their termination ability removed, this means that once your monomer has been consumed, the addition of more monomer will result in the polymer chains continuing to grow until all of the additional monomer is consumed. This will continue until the metal catalyst at the end of the chain is intentionally removed by the addition of a quenching agent. As a result, it may potentially allow one to create a block or gradient copolymer fairly easily and accurately. This can lead to a high ability to tune the properties of the polymer to a desired application (electrical/ionic conduction, etc.).

"Living" free Radical Polymerization

Starting in the 1970s several new methods were discovered which allowed the development of living polymerization using free radical chemistry. These techniques involved catalytic chain transfer polymerization, iniferter mediated polymerization, stable free radical mediated polymerization (SFRP), atom transfer radical polymerization (ATRP), reversible addition-fragmentation chain transfer (RAFT) polymerization, and iodine-transfer polymerization.

In "living" radical polymerization (or controlled radical polymerization (CRP)) the chain breaking pathways are severely depressed when compared to conventional radical polymerization (RP) and CRP can display characteristics of a living polymerization. However, since chain termination is not absent, but only minimized, CRP technically does not meet the requirements imposed by IUPAC for a living polymerization.

There are two general strategies employed in CRP to suppress chain breaking reactions and promote fast initiation relative to propagation. Both strategies are based on developing a dynamic equilibrium amongst an active propagating radical and a dormant species.

The first strategy involves a reversible trapping mechanism in which the propagating radical undergoes an activation/deactivation (i.e. Atom-transfer radical-polymerization) process with a species X. The species X is a persistent radical, or a species that can generate a stable radical, that cannot terminate with itself or propagate but can only reversibly "terminate" with the propagating radical (from the propagating polymer chain)P*. P* is a radical species that an propagate (k_p) and irreversibly terminate (k_t) with another P*. X is normally a nitroxide (i.e. TEMPO used in Nitroxide Mediated Radical Polymerization) or an organometallic species. The dormant species (P_n-X) can be activated to regenerate the active propagating species (P*) spontaneously, thermally, using a catalyst and optically.

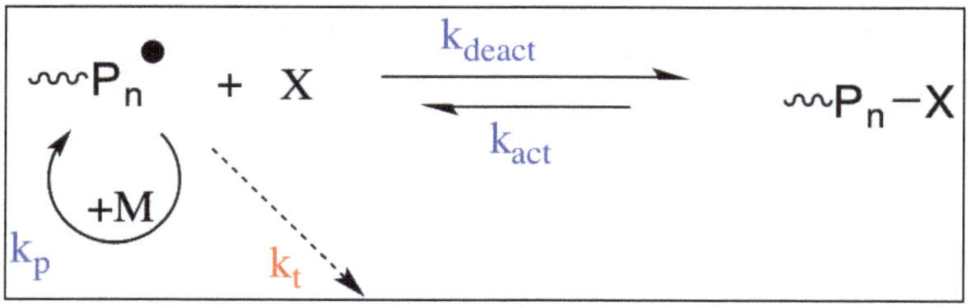

The second strategy is based on a degenerative transfer (DT) of the propagating radical between transfer agent that acts as a dormant species (i.e. Reversible addition–fragmentation chain-transfer polymerization). The DT based CRP's follow the conventional kinetics of radical polymerization, that is slow initiation and fast termination, but the transfer agent (Pm-X or Pn-X) is present in a much higher concentration compared to the radical initiator.The propagating radical species undergoes a thermally neutral exchange with the dormant transfer agent through atom transfer, group transfer or addition fragment chemistry.

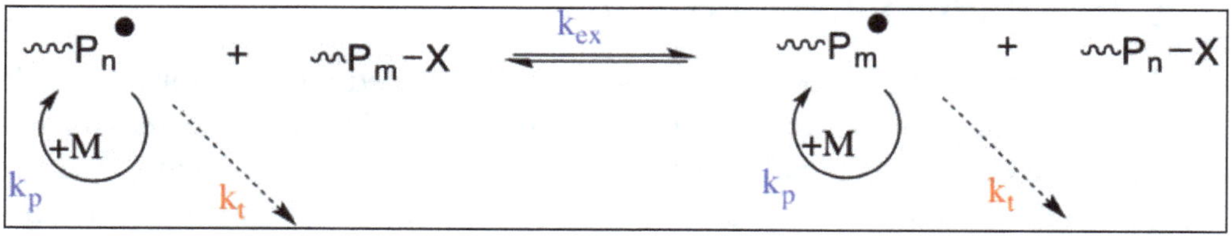

Living Chain-growth Polycondensations

Chain growth polycondensation polymerizations were initially developed under the premise that a change in substituent effects of the polymer, relative to the monomer, causes the polymers end group to be more reactive this has been referred to as "reactive intermediate polycondensation". The essential result is monomers preferentially react with the activated polymer end groups over reactions with other monomers. This preferred reactivity is the fundamental difference when categorizing a polymerization mechanism as chain-growth as opposed to step-growth in which the monomer and polymer chain end group have equal reactivity (the reactivity is uncontrolled). Several strategies were employed to minimize monomer-monomer reactions (or self-condensation) and polymerizations with low D and controllable Mn have been attained by this mechanism for small molecular weight polymers. However, for high molecular weight polymer chains (i.e. small initiator to monomer ratio) the Mn is not easily to controlled, for some monomers, since self-condensation between monomers occurred more frequently due to the low propagating species concentration.

Catalyst-transfer Polycondensation

Catalyst transfer polycondensation (CTP) is a chain-growth polycondensation mechanism in which the monomers do not directly react with one another and instead the monomer will only react with the polymer end group through a catalyst-mediated mechanism. The general process consists of the catalyst activating the polymer end group followed by a reaction of the end group with a 2nd incoming monomer. The catalyst is then transferred to the elongated chain while activating the end group (as shown below).

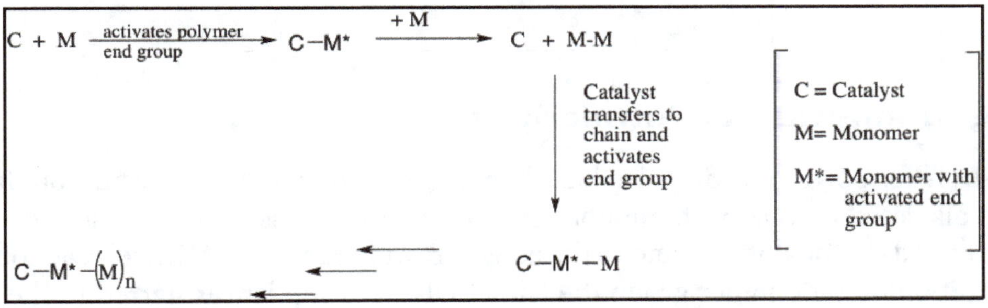

Catalyst transfer polycondensation allows for the living polymerization of π-conjugated polymers and was discovered by Tsutomu Yokozawa in 2004 and Richard McCullough. In CTP the propagation step is based on organic cross coupling reactions (i.e. Kumada coupling, Sonogashira coupling, Negishi coupling) top form carbon carbon bonds between difunctional monomers. When Yokozawa and McCullough independently discovered the polymerization using a metal catalyst to couple a Grignard reagent with an organohalide making a new carbon-carbon bond. The mechanism below shows the formation of poly(3-alkylthiophene) using a Ni initiator (L_n can be 1,3-Bis(diphenylphosphino)propane (dppp)) and is similar to the conventional mechanism for Kumada coupling involving an oxidative addition, a transmetalation and a reductive elimination step. However, there is a key difference, following reductive elimination in CTP, an associative complex is formed (which has been supported by intra-/intermolecular oxidative addition competition experiments) and the subsequent oxidative addition occurs between the metal center and the associated chain (an intramolecular pathway). Whereas in a coupling reaction the newly formed alkyl/aryl compound diffuses away and the subsequent oxidative addition occurs between an incoming Ar-Br bond and the metal center. The associative complex is essential to for polymerization to occur in a living fashion since it allows the metal to undergo a preferred intramolecular oxidative addition and remain with a single propagating chain (consistent with chain-growth mechanism), as opposed to an intermolecular oxidative addition with other monomers present in the solution (consistent with a step-growth, non-living, mechanism). The monomer scope of CTP has been increasing since its discovery and has included poly(phenylene)s, poly(fluorine)s, poly(selenophene)s and poly(pyrrole)s.

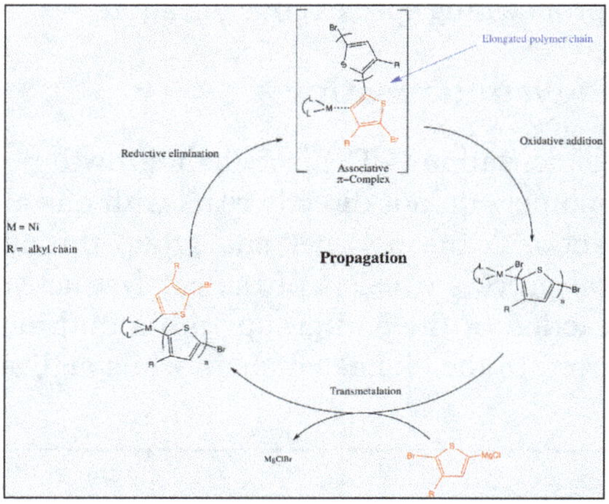

Living Group-transfer Polymerization

Group-transfer polymerization also has characteristics of living polymerization. It is applied to alkylated methacrylate monomers and the initiator is a silyl ketene acetal. New monomer adds to the initiator and to the active growing chain in a Michael reaction. With each addition of a monomer group the trimethylsilyl group is transferred to the end of the chain. The active chain-end is not ionic as in anionic or cationic polymeriation but is covalent. The reaction can be catalysed by bifluorides and bioxyanions such as *tris(dialkylamino)sulfonium bifluoride* or *tetrabutyl ammonium bibenzoate*. The method was discovered in 1983 by O.W. Webster and the name first suggested by Barry Trost.

Applications

Living polymerizations can be (and in some cases are) used industrially for many different applications. They can range from self-healing materials for space equipment to the easy design of copolymers for ion-exchange membranes in fuel cells, nanoscale lithography, etc.. While living polymerizations are still not widely used industrially, the field is rapidly growing, as well as the list of practical applications.

Self-healing Materials

Self-healing materials are materials in which repair, or "heal", themselves upon damage from an external force through the use of living polymers. For example, if a crack forms in the material, it proceeds to repair the crack and restore itself to its original, undamaged form. It achieves this by incorporating monomer-containing beads into a material made of a living polymer (with a terminally active chain). This has been achieved recently using a polyurethane derivative, with beads of monomer embedded in the material that become opened upon cracking of the material.

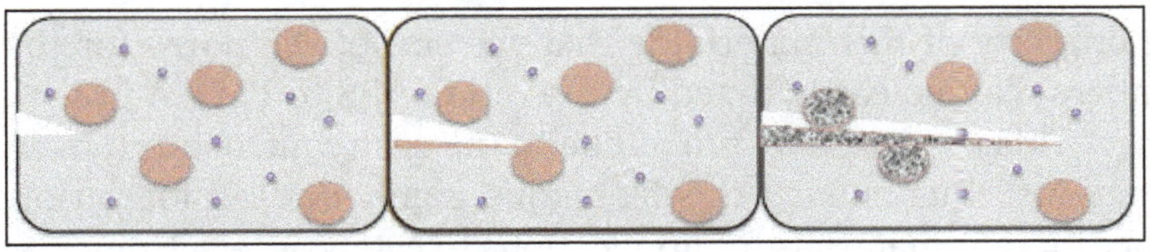

The polymer that makes up the material is designed as a living polymer, with reactive terminal end-groups that bind to the freshly provided monomer upon damage to the microbeads. This addition of monomer to the polymer chain increases the polymer chain to a length that fills the once open crack, in essence reconnecting all of the pieces back into one. According to Odriozola and coworkers, this application is originally designed for space equipment (in the event of debris damaging the equipment).

Copolymer Synthesis and Applications

Copolymers are polymers consisting of multiple different monomer species, and can be arranged in various orders, three of which are seen in the figure below.

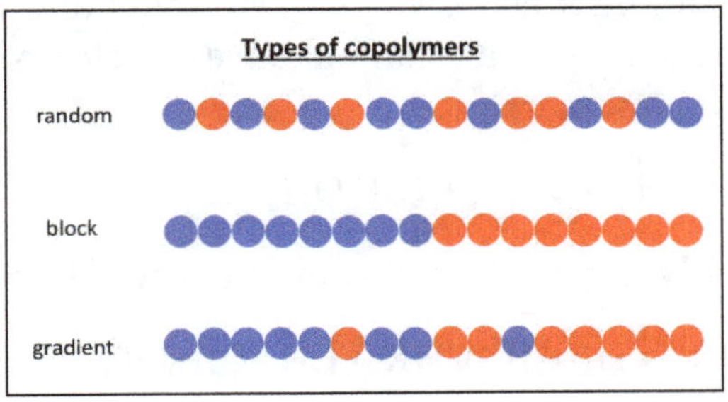

While there exist others (alternating copolymers, graft copolymers, and stereoblock copolymers), these three are more common in the scientific literature. In addition, block copolymers can exist as many types, including triblock (A-B-A), alternating block (A-B-A-B-A-B), etc.

Of these three types, block and gradient copolymers are commonly synthesized through living polymerizations, due to the ease of control living polymerization provides. Copolymers are highly desired due to the increased flexibility of properties a polymer can have compared to their homopolymer counterparts. The synthetic techniques used range from ROMP to generic anionic or cationic living polymerizations.

Copolymers, due to their unique tunability of properties, can have a wide range of applications. One example (of many) is nano-scale lithography using block copolymers. One used frequently is a block copolymer made of polystyrene and poly(methyl methacrylate) (abbreviated PS-*b*-PMMA). This copolymer, upon proper thermal and processing conditions, can form cylinders on the order of a few tens of nanometers in diameter of PMMA, surrounded by a PS matrix. These cylinders can then be etched away under high exposure to UV light and acetic acid, leaving a porous PS matrix.

The unique property of this material is that the size of the pores (or the size of the PMMA cylinders) can be easily tuned by the ratio of PS to PMMA in the synthesis of the copolymer. This can be easily tuned due to the easy control given by living polymerization reactions, thus making this technique highly desired for various nanoscale patterning of different materials for applications to catalysis, electronics, etc.

Cationic Polymerization

Many vinyl monomers readily polymerize in the presence of very small amounts of catalyst of the type used in Friedel-Crafts reactions. Examples of effective catalysts are $AlCl_3$, $AlBr_3$, BF_3, $TiCl_4$, $SnCl_4$, and in some cases strong acids such as H_2SO_4. All these catalysts are examples of Lewis acids with strong electron-acceptor capability. They usually require a co-catalyst, namely a Lewis base such as Water, acetic acid or alcohol:

$BF_3 + H_2O \Leftrightarrow H^+BF_3OH^-$

Monomers that polymerize in the presence of these catalysts include isobutylene, styrene, alpha-methylstyrene, butadiene, vinyl alkyl ethers and many other monomers having electron-donating substituents that enhance the electron-sharing ability of the double bond of the vinyl monomers.

$H^+BF_3OH^- + CH_2{=}CHR \rightarrow H_3C\text{-}C^+HR + (BF_3OH)^-$

They all can be readily polymerized to high-molecular weight polymers.

$H_3C\text{-}C^+HR + n\ CH_2{=}CHR \rightarrow H(\text{-}CH_2\text{-}CHR\text{-})_nCH_2\text{-}C^+HR$

However, some other monomers, such as propylene and other olefins, reach only low to medium molecular weights when polymerized with strong Lewis acids.

The cationic polymerization usually proceeds at high rates both at high and (very) low temperatures. For this reason, a uniform and constant reaction condition cannot be maintained during polymerization. For example, isobutylene at -100C in the presence of a strong Lewis acid polymerizes to high molecular weight polybutylene within a fraction of a second. To prevent excessive rise in temperature in the reaction vessel, an "internal refrigerant" is usually added to the mixture that dissipates the heat by evaporation of a portion of the liquid.

Both the rate of reaction and the molecular weight decreases with increasing temperature. For this reason, low temperatures are usually preferred. In fact, molecular weights obtained at room temperature are often much lower than those achieved by free-radical polymerization.

Synthesis

Initiation

Initiation is the first step in cationic polymerization. During initiation, a carbenium ion is generated from which the polymer chain is made. The counterion should be non-nucleophilic, otherwise the reaction is terminated instantaneously. There are a variety of initiators available for cationic polymerization, and some of them require a coinitiator to generate the needed cationic species.

Classical Protonic Acids

Strong protic acids can be used to form a cationic initiating species. High concentrations of the acid are needed in order to produce sufficient quantities of the cationic species. The counterion (A^-) produced must be weakly nucleophilic so as to prevent early termination due to combination with the protonated olefin. Common acids used are phosphoric, sulfuric, fluro-, and triflic acids. Only low molecular weight polymers are formed with these initiators.

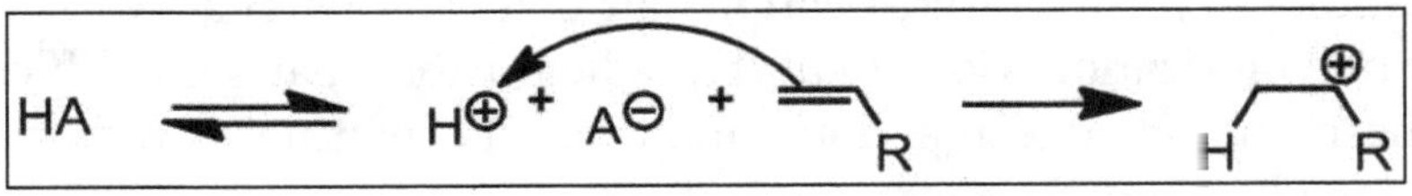

Initiation by protic acids.

Lewis acids/Friedel-Crafts catalysts

Lewis acids are the most common compounds used for initiation of cationic polymerization. The more popular Lewis acids are $SnCl_4$, $AlCl_3$, BF_3, and $TiCl_4$. Although these Lewis acids alone are able to induce polymerization, the reaction occurs much faster with a suitable cation source. The cation source can be water, alcohols, or even a carbocation donor such as an ester or an anhydride. In these systems the Lewis acid is referred to as a coinitiator while the cation source is the initiator. Upon reaction of the initiator with the coinitiator, an intermediate complex is formed which then goes on to react with the monomer unit. The counterion produced by the initiator-coinitiator complex is less nucleophilic than that of the Brønsted acid A^- counterion. Halogens, such as chlorine and bromine, can also initiate cationic polymerization upon addition of the more active Lewis acids.

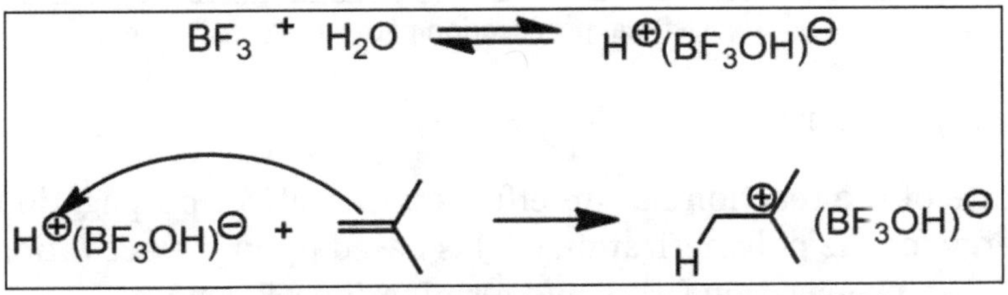

Initiation with boron trifluoride (coinitiator) and water (initiator)

Carbenium ion salts

Stable carbenium ions are used to initiate chain growth of only the most reactive olefins and are known to give well defined structures. These initiators are most often used in kinetic studies due to the ease of being able to measure the disappearance of the carbenium ion absorbance. Common carbenium ions are trityl and tropylium cations.

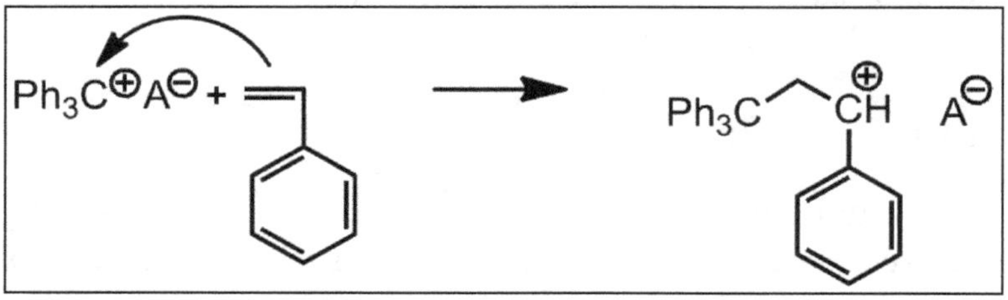

Initiation with trityl carbenium ion.

Ionizing Radiation

Ionizing radiation can form a radical-cation pair that can then react with a monomer to start cationic polymerization. Control of the radical-cation pairs are difficult and often depend on the monomer and reaction conditions. Formation of radical and anionic species are often observed.

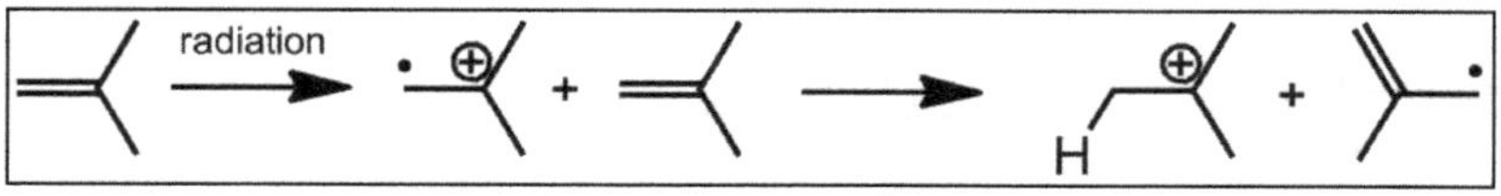

Initiation using ionizing radiation.

Propagation

Propagation proceeds via addition of monomer to the active species, i.e. the carbenium ion. The monomer is added to the growing chain in a head-to-tail fashion; in the process, the cationic end group is regenerated to allow for the next round of monomer addition.

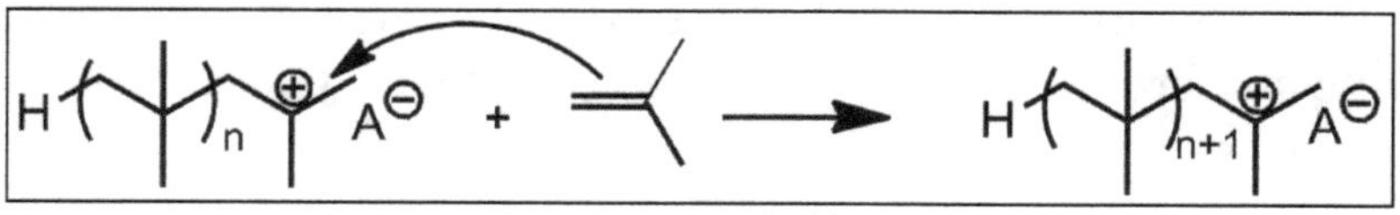

General propagation pathway.

Effect of Temperature

The temperature of the reaction has an effect on the rate of propagation. The overall activation energy for the polymerization (E) is based upon the activation energies for the initiation (E_i), propagation (E_p), and termination (E_t) steps:

$$E = E_i + E_p - E_t$$

Generally, E_t is larger than the sum of E_i and E_p, meaning the overall activation energy is negative. When this is the case, a decrease in temperature leads to an increase in the rate of propagation. The converse is true when the overall activation energy is positive.

Chain length is also affected by temperature. Low reaction temperatures, in the range of 170–190 K, are preferred for producing longer chains. This comes as a result of the activation energy for termination and other side reactions being larger than the activation energy for propagation. As the temperature is raised, the energy barrier for the termination reaction is overcome, causing shorter chains to be produced during the polymerization process.

Effect of Solvent and Counterion

The solvent and the counterion (the gegen ion) have a significant effect on the rate of propagation. The counterion and the carbenium ion can have different associations according to intimate ion pair theory; ranging from a covalent bond, tight ion pair (unseparated), solvent-separated ion pair (partially separated), and free ions (completely dissociated).

~~~RXcovalent ~~~$R^+X^{-\text{tight ion pair}}$ ~~~$R^+/X^{-\text{solvent-separatedion pair}}$ ~~~$R^+ + X^{-\text{free ions}}$

Range of associations between the carbenium ion (R^+) and gegen ion (X^-)

The association is strongest as a covalent bond and weakest when the pair exists as free ions. In cationic polymerization, the ions tend to be in equilibrium between an ion pair (either tight or solvent-separated) and free ions. The more polar the solvent used in the reaction, the better the solvation and separation of the ions. Since free ions are more reactive than ion pairs, the rate of propagation is faster in more polar solvents.

The size of the counterion is also a factor. A smaller counterion with a higher charge density, will have stronger electrostatic interactions with the carbenium ion than will a larger counterion which has a lower charge density. Further, a smaller counterion is more easily solvated by a polar solvent than a counterion with low charge density. The result is increased propagation rate with increased solvating capability of the solvent.

Termination

Termination generally occurs via unimolecular rearrangement with the counterion. In this process, an anionic fragment of the counterion combines with the propagating chain end. This not only inactivates the growing chain, but it also terminates the kinetic chain by reducing the concentration of the initiator-coinitiator complex.
~~~

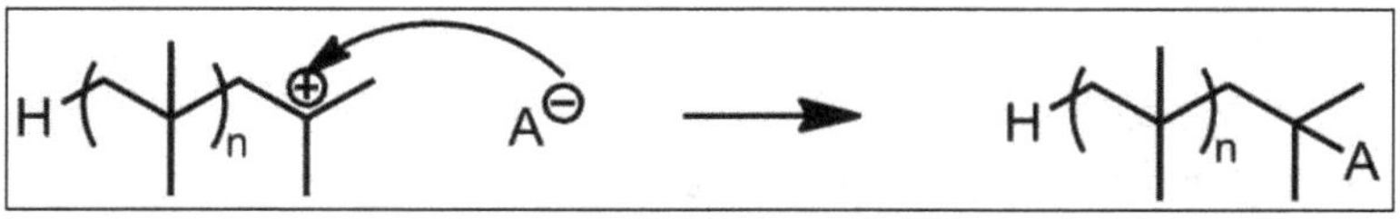

Termination by combination with an anionic fragment from the counterion.

Chain Transfer

Chain transfer can take place in two ways. One method of chain transfer is hydrogen abstraction from the active chain end to the counterion. In this process, the growing chain is terminated, but the initiator-coinitiator complex is regenerated to initiate more chains.

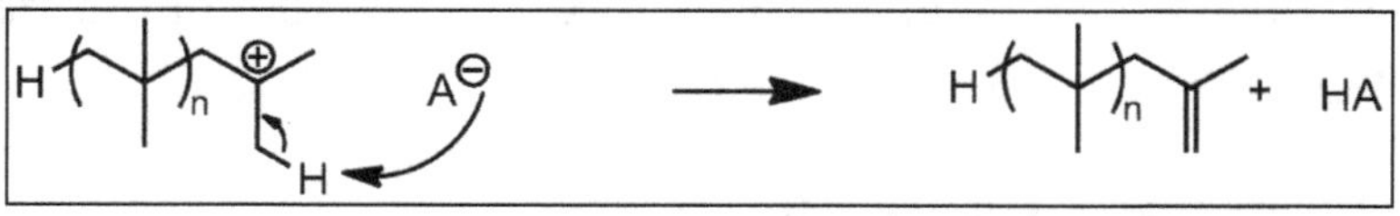

Chain transfer by hydrogen abstraction to the counterion.

The second method involves hydrogen abstraction from the active chain end to the monomer. This terminates the growing chain and also forms a new active carbenium ion-counterion complex which can continue to propagate, thus keeping the kinetic chain intact.

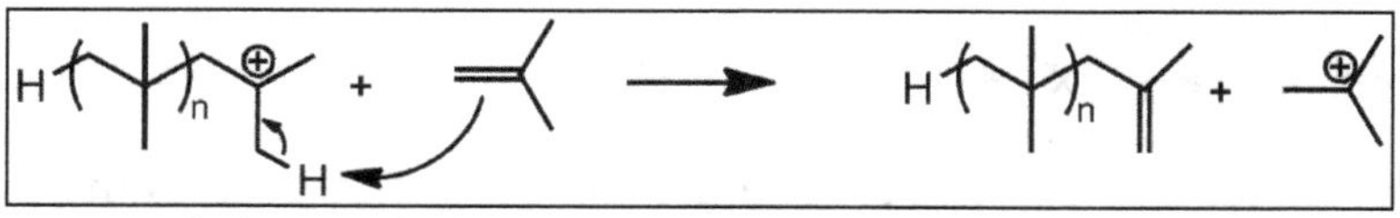

Chain transfer by hydrogen abstraction to the monomer.

Cationic Ring-opening Polymerization

Cationic ring-opening polymerization follows the same mechanistic steps of initiation, propagation, and termination. However, in this polymerization reaction, the monomer units are cyclic in comparison to the resulting polymer chains which are linear. The linear polymers produced can have low ceiling temperatures, hence end-capping of the polymer chains is often necessary to prevent depolymerization.

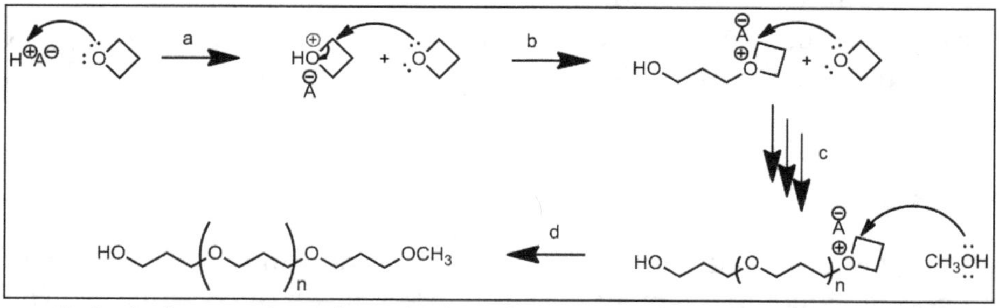

Cationic ring-opening polymerization of oxetane involving (a and b) initiation, (c) propagation, and (d) termination with methanol.

Kinetics

The rate of propagation and the degree of polymerization can be determined from an analysis of the kinetics of the polymerization. The reaction equations for initiation, propagation, termination, and chain transfer can be written in a general form:

$$\mathrm{I}{+}{+}\mathrm{M} \xrightarrow{k_i} \mathrm{M}^+$$

$$\mathrm{M}^+ + \mathrm{M} \xrightarrow{k_p} \mathrm{M}^+$$

$$\mathrm{M}^+ \xrightarrow{k_t} \mathrm{M}$$

$$\mathrm{M}^+ + \mathrm{M} \xrightarrow{k_{tr}} \mathrm{M} + \mathrm{M}^+$$

In which I^+ is the initiator, M is the monomer, M^+ is the propagating center, and k_i, k_p, k_t, and k_{tr} are the rate constants for initiation, propagation, termination, and chain transfer, respectively. For simplicity, counterions are not shown in the above reaction equations and only chain transfer to monomer is considered. The resulting rate equations are as follows, where brackets denote concentrations:

$$\text{rate(initiation)} = k_i[\mathrm{I}^+][\mathrm{M}]$$
$$\text{rate(propagation)} = k_p[\mathrm{M}^+][\mathrm{M}]$$
$$\text{rate(termination)} = k_t[\mathrm{M}^+]$$
$$\text{rate(chain transfer)} = k_{tr}[\mathrm{M}+][\mathrm{M}]$$

Assuming steady-state conditions, i.e. the rate of initiation = rate of termination:

$$[\mathrm{M}^+] = \frac{k_i[I^+][\mathrm{M}]}{k_t}$$

This equation for $[M^+]$ can then be used in the equation for the rate of propagation:

$$\text{rate(propagation)} = \frac{k_p k_i [\mathrm{M}]^2[\mathrm{I}^+]}{k_t}$$

From this equation, it is seen that propagation rate increases with increasing monomer and initiator concentration.

The degree of polymerization, X_n, can be determined from the rates of propagation and termination:

$$X_n = \frac{\text{rate(propagation)}}{\text{rate(termination)}} = \frac{k_p[\mathrm{M}]}{k_t}$$

If chain transfer rather than termination is dominant, the equation for X_n becomes

$$X_n = \frac{\text{rate(propagation)}}{\text{rate(chain transfer)}} = \frac{k_p}{k_{tr}}$$

Living Polymerization

In 1984, Higashimura and Sawamoto reported the first living cationic polymerization for alkyl vinyl ethers. This type of polymerization has allowed for the control of well-defined polymers. A key characteristic of living cationic polymerization is that termination is essentially eliminated, thus the cationic chain growth continues until all monomer is consumed.

Commercial Applications

The largest commercial application of cationic polymerization is in the production of polyisobutylene (PIB) products which include polybutene and butyl rubber. These polymers have a variety of applications from adhesives and sealants to protective gloves and pharmaceutical stoppers. The reaction conditions for the synthesis of each type of isobutylene product vary depending on the desired molecular weight and what type(s) of monomer(s) is used. The conditions most commonly used to form low molecular weight (5–10 x 10^4 Da) polyisobutylene are initiation with $AlCl_3$, BF_3, or $TiCl_4$ at a temperature range of −40 to 10 °C. These low molecular weight polyisobutylene polymers are used for caulking and as sealants. High molecular weight PIBs are synthesized at much lower temperatures of −100 to −90 °C and in a polar medium of methylene chloride. These polymers are used to make uncrosslinked rubber products and are additives for certain thermoplasts. Another characteristic of high molecular weight PIB is its low toxicity which allows it to be used as a base for chewing gum. The main chemical companies that produce polyisobutylene are Esso, ExxonMobil, and BASF.

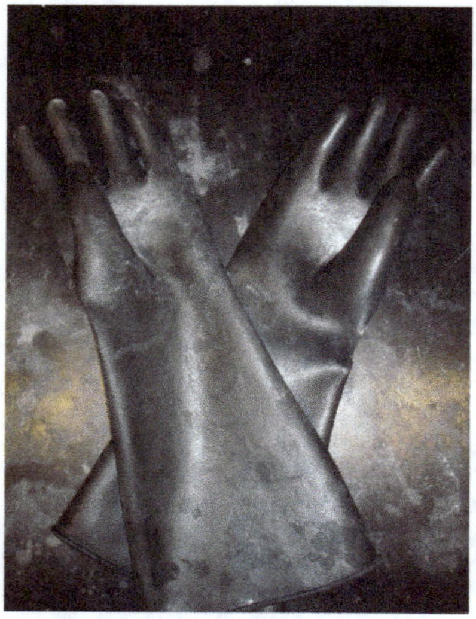

Butyl rubber gloves.

Butyl rubber, in contrast to PIB, is a copolymer in which the monomers isobutylene (~98%) and isoprene (2%) are polymerized in a process similar to high molecular weight PIBs. Butyl rubber polymerization is carried out as a continuous process with $AlCl_3$ as the initiator. Its low gas permeability and good resistance to chemicals and aging make it useful for a variety of applications such as protective gloves, electrical cable insulation, and even basketballs. Large scale production of butyl rubber started during World War II, and roughly 1 billion pounds/year are produced in the U.S. today.

Polybutene is another copolymer, containing roughly 80% isobutylene and 20% other butenes (usually 1-butene). The production of these low molecular weight polymers (300–2500 Da) is done within a large range of temperatures (−45 to 80 °C) with $AlCl_3$ or BF_3. Depending on the molecular weight of these polymers, they can be used as adhesives, sealants, plasticizers, additives for transmission fluids, and a variety of other applications. These materials are low-cost and are made by a variety of different companies including BP Chemicals, Esso, and BASF.

Other polymers formed by cationic polymerization are homopolymers and copolymers of polyterpenes, such as pinenes (plant-derived products), that are used as tackyfiers. In the field of heterocycles, 1,3,5-trioxane is copolymerized with small amounts of ethylene oxide to form the highly crystalline polyoxymethylene plastic. Also, the homopolymerization of alkyl vinyl ethers is achieved only by cationic polymerization.

Anionic Polymerization

Addition polymers can also be made by chain reactions that proceed through intermediates that carry either a negative or positive charge.

When the chain reaction is initiated and carried by negatively charged intermediates, the reaction is known as anionic polymerization. Like free-radical polymerizations, these chain reactions take place via chain-initiation, chain-propagation, and chain-termination steps.

Monomer Characteristics

In order for polymerization to occur with vinyl monomers, the substituents on the double bond must be able to stabilize a negative charge. Stabilization occurs through delocalization of the negative charge. Because of the nature of the carbanion propagating center, substituents that react with bases or nucleophiles either must not be present or be protected.

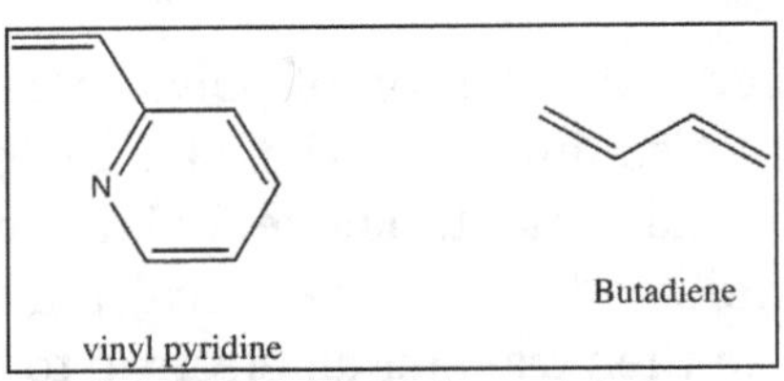

Examples of vinyl monomers.

Vinyl monomers with substituents that stabilize the negative charge through charge delocalization, undergo polymerization without termination or chain transfer. These monomers include styrene, dienes, methacrylate, vinyl pyridine, aldehydes, epoxide, episulfide, cyclic siloxane, and lactones. Polar monomers, using controlled conditions and low temperatures, can undergo anionic polymerization. However, at higher temperatures they do not produce living stable, carbanionic chain ends because their polar substituents can undergo side reactions with both initiators and propagating chain centers. The effects of counterion, solvent, temperature, Lewis base additives, and inorganic solvents have been investigated to increase the potential of anionic polymerizations of polar monomers. Polar monomers include acrylonitrile, cyanoacrylate, propylene oxide, vinyl ketone, acrolein, vinyl sulfone, vinyl sulfoxide, vinyl silane and isocyanate.

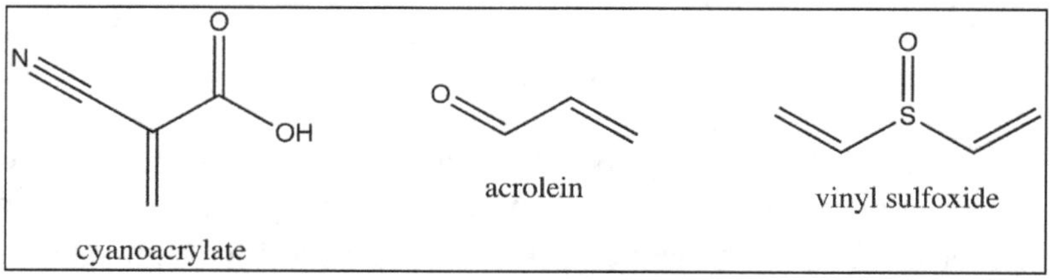

Examples of polar monomers.

Solvent

The solvent used in anionic addition polymerizations are determined by the reactivity of both the initiator and carbanion of the propagating chain end. Anionic species with low reactivity, such as heterocyclic monomers, can use a wide range of solvents.

Initiation

The reactivity of initiators used in anionic polymerization should be similar to that of the monomer that is the propagating species. The pKa values for the conjugate acids of the carbanions formed from monomers can be used to deduce the reactivity of the monomer. The least reactive monomers have the largest pKa values for their corresponding conjugate acid and thus, require the most reactive initiator. Two main initiation pathways involve electron transfer (through alkali metals) and strong anions.

Initiation by Electron Transfer

Szwarc and coworkers studied the initiation of polymerization through the use of aromatic radical-anions such as sodium naphthenate. In this reaction, an electron is transferred from the alkali metal to naphthalene. Polar solvents are necessary for this type of initiation both for stability of the anion-radical and to solvate the cation species formed. The anion-radical can then transfer an electron to the monomer.

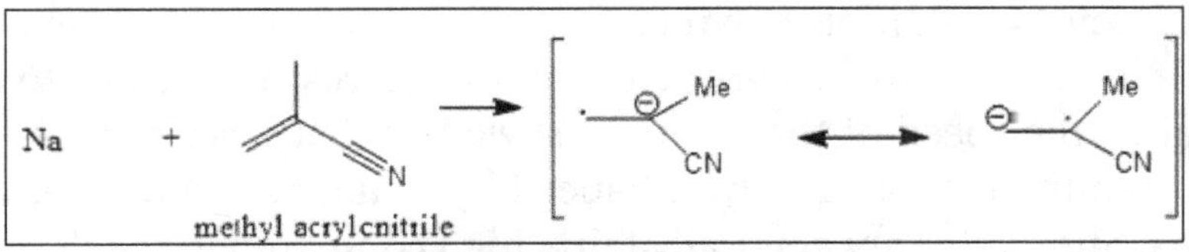

Initiation through electron transfer.

Initiation can also involve the transfer of an electron from the alkali metal to the monomer to form an anion-radical. Initiation occurs on the surface of the metal, with the reversible transfer of an electron to the adsorbed monomer.

Initiation by Strong Anions

Nucleophilic initiators include covalent or ionic metal amides, alkoxides, hydroxides, cyanides, phosphines, amines and organometallic compounds (alkyllithium compounds and Grignard reagents). The initiation process involves the addition of a neutral (B:) or negative (B:-) nucleophile to the monomer.

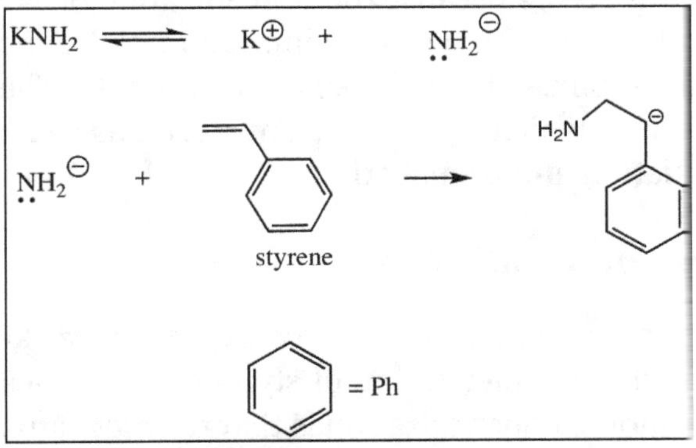

Initiation through strong anion.

The most commercially useful of these initiators has been the alkyllithium initiators. They are primarily used for the polymerization of styrenes and dienes.

Monomers activated by strong electronegative groups may be initiated even by weak anionic or neutral nucleophiles (i.e. amines, phosphines). Most prominent example is the curing of cyanoacrylate, which constitutes the basis for superglue. Here, only traces of basic impurities are sufficient to induce an anionic addition polymerization or zwitterionic addition polymerization, respectively.

Propagation

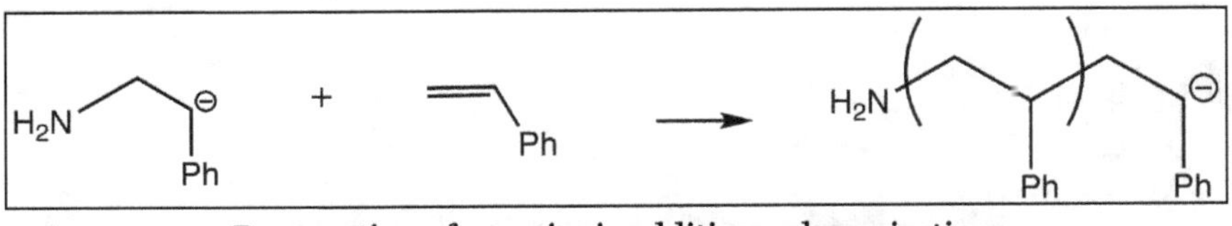

Propagation of an anionic addition polymerization.

Propagation in anionic addition polymerization results in the complete consumption of monomer. It is very fast and occurs at low temperatures. This is due to the anion not being very stable, the speed of the reaction as well as that heat is released during the reaction. The stability can be greatly enhanced by reducing the temperatures to near 0 °C. The propagation rates are generally fairly high compared to the decay reaction, so the overall polymerization rates is generally not affected.

Termination

Anionic addition polymerizations have no formal termination pathways because proton transfer from solvent or other positive species does not occur. However, termination can occur through unintentional quenching due to trace impurities. This includes trace amounts of oxygen, carbon dioxide or water. Intentional termination can occur through the addition of water or alcohol. Another method of termination, chain transfer, can occur when an agent can act as a Brønsted acid. In this case, the pKa value of the agent is similar to the conjugate acid of the propagating carbanionic chain end. Spontaneous termination occurs because the concentration of carbanion centers decay over time and eventually results in hydride elimination. Polar monomers are more reactive because they are stabilized by their polar substituents. These polar substituents can react with nucleophiles which results in termination as well as side reactions that compete with both initiation and propagation.

Living Anionic Polymerization

Living polymerization was demonstrated by Szwarc and co workers in 1956. Their initial work was based on the polymerization of styrene and dienes. One of the remarkable features of living anionic polymerization is that the mechanism involves no formal termination step. In the absence of impurities, the carbanion would still be active and capable of adding another monomer. The chains will remain active indefinitely unless there is inadvertent or deliberate termination or chain transfer.

Kinetics

The kinetics of anionic addition polymerization depend on whether or not a termination pathway occurs.

Kinetics of Living Anionic Addition Polymerization

In general, the reaction mechanism for living anionic addition polymerization are as follows:

$$I^- + M \xrightarrow{k_{init}} M^-$$

$$M^- + M \xrightarrow{k_{prop}} M^-$$

where I = initiator, k_{init} = the initiation reaction rate constant, M = monomer, M^-= propagating species, and k_{prop} = the propagation reaction rate constant.

As most polymerizations of this type do not have a termination pathway, the rate of polymerization is the rate of propagation:

$$\text{rate(prop)} = k_p[\text{M}^-][\text{M}]$$

where k_p is the rate of constant of propagation, $[M^-]$ is the total concentration of propagating centers, and [M] is the concentration of monomer. Since there is no termination pathway in living polymerizations, the concentration of propagating centers is equal to the concentration of initiator. Thus,

$$\text{rate(prop)} = k_p[\text{I}][\text{M}]$$

The degree of polymerization, X_n is also affected by no termination pathway. It is the ratio of concentration of reacted monomer ($[M]_o$) to initiator($[I]_o$) times the percent conversion *p*. In this case, the chain length (v) is equal to X_n.

$$v = \frac{[\text{M}]_o}{[\text{I}]_o}\rho$$

When conversion, $p = 1$ (100% conversion), chain length is simply the ratio of reacted monomer to initiator.

$$v = \frac{[\text{M}]_o}{[\text{I}]_o}$$

Kinetics: termination Due to Impurities

When termination occurs due to impurities, the impurities must be taken into account in determining the reaction rate. The reaction mechanisms would begin the same as that of a living anionic addition (initiation and propagation). However, there would now be a termination step to account for the effect of the impurities on the reaction.

$$M^- + HX \xrightarrow{k_{term}} M - H + X^-$$

where M^-= propagating species, HX = impurity and k_{term} = the termination reaction rate constant.

Using the steady-state approximation, the rate of propagation becomes

$$\text{rate(prop)} = \frac{k_{init}k_{prop}[\text{I}][\text{M}]^2}{k_{term}[\text{H-X}]}$$

Since ,

$$\nu = \frac{\text{rate(prop)}}{\text{rate(term)}} = \frac{k_{prop}[\text{M}]}{k_{term}[\text{H-X}]}$$

Thus chain length and rate of propagation are negatively impacted by the presence of impurities in the reaction.

Ring-opening Polymerization

A ring-opening polymerization (ROP) is another form of chain-growth polymerization in which the terminal end group of a polymer chain acts as a reactive center where further cyclic monomers can be added by ring-opening and additon of the broken bond. Typical cyclic monomers that can be polymerized via ROP are di-functional monomers that carry two different reactive groups like one amine or alcohol and one carboxylic acid that have undergone a cyclization reaction. Two examples are caprolactam and caprolactone:

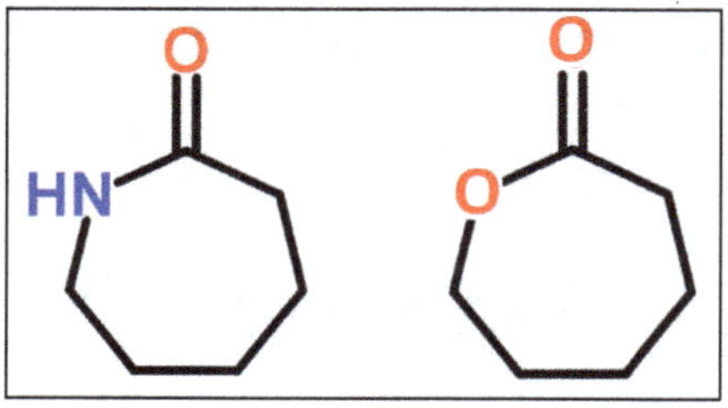

To polymerize these moieties, one of the rings has to open prior polymerization. This can be achieved, for example, by adding a small amount of a nucleophilic reagent (Lewis base) as an initiator. This reaction is called *anionic ring-opening polymerization* (AROP):

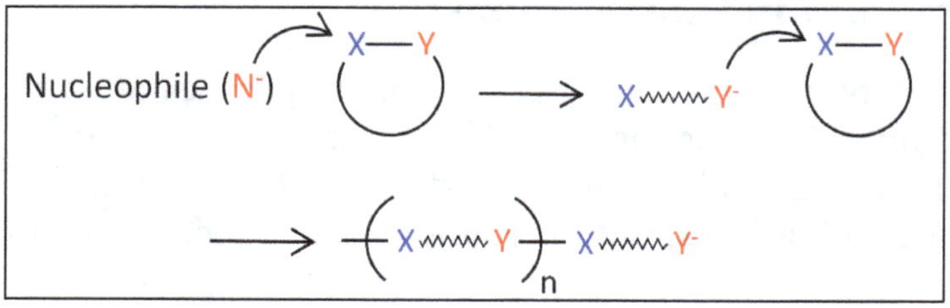

Two well known thermoplastic polymers that can be synthesized via anionic ring-opening polymerization are polycaprolactam (Nylon 6) and polycaprolactone (PCL):

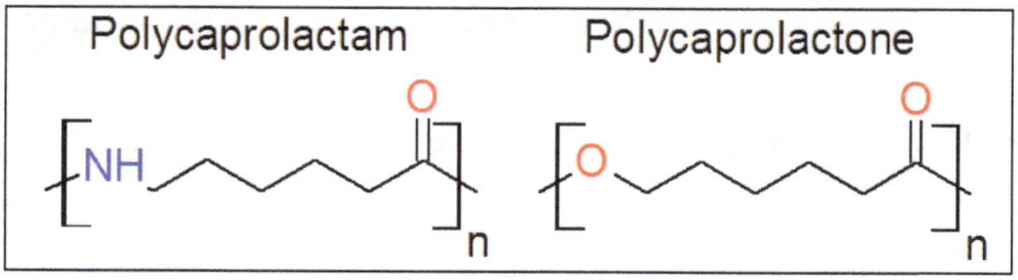

Most monomers that undergo AROP contain polar bonds like ester, amide, carbonate, urethane, epoxide, and phosphate which polymerize to the corresponding polyester, polyamide, polycarbonate, polyurethane, polyepoxide, and polyphosphate.

Cationic ring-opening polymerization (CROP) is also possible In this case, a small amount of an electrophilic reagent (Lewis acid) is added to the monomer to initiate polymerization. However, not all cyclic monomers containing an heteroatom undergo CROP. Whether and how readily a cyclic monomer undergoes CROP depends on the ring size, to be more specific, on the ring strain. Cyclic monomers with small or no ring strain will not polymerize whereas small rings with greater ring strain like 4, 6, and 7-membered rings of cyclic esters, polymerize readily through CROP.

Some examples of cyclic monomers that polymerize through anionic or cationic ring-opening polymerization include cyclic ethers, lactones, lactams, and epoxides.

Ring-opening polymerization can also proceed via *free radical polymerization*. The introduction of an oxygen into the ring will usually promote free radical ring-opening polymerization, because the resulting carbon–oxygen double bond is much more stable than a carbon-carbon double bond. Thus, cyclic hetero monomers that carry a vinyl side group like cyclic ketene acetals, cyclic ketene aminals, cyclic vinyl ethers, and unsaturated spiro ortho esters will readily undergo free radical ring-opening polymerization. Copolymerization of these monomers with a wide variety of vinyl monomers will introduce ester, amide, keto or carbonate groups into the backbone, which results in functionally terminated oligomers.

References

- Odian, George (27 Feb 2004). “8”. Principles of Polymerization (4th ed.). John Wiley & Sons, Inc. p. 633. ISBN 978-0-471-27400-1
- Cowie, J. M. G. Polymers: Chemistry and Physics of Modern Materials (2nd ed.). Blackie (USA: Chapman & Hall). pp. 58–60. ISBN 0-216-92980-6
- Living-polymerization-and-molecular-weight: polymersolutions.com, Retrieved 2, May 2020
- “Stable Free Radical Polymerization”. Xerox Corp. 2010. Archived from the original on 28 November 2003. Retrieved 10, March 2020
- Islamova, R. M.; Puzin, Y. I.; Kraikin, V. A.; Fatykhov, A. A.; Dzhemilev, U. M. (2006). “Controlling the Polymerization of Methyl Methacrylate with Ternary Initiating Systems”. Russian Journal of Applied Chemistry. 79 (9): 1509–1513. doi:10.1134/S1070427206090229
- Polymerization: adhesiveandglue.com, Retrieved 20, July 2020
- “Polymer Synthesis”. Case Western Reserve University. 2009. Archived from the original on 7 February 2010. Retrieved 21, March 2020
- Alfrey, Turner; Price, Charles C. “Relative reactivities in vinyl copolymerization”. Journal of Polymer Science. 2 (1): 101–106. Bibcode:1947JPoSc...2..101A. doi:10.1002/pol.1947.120020112
- Macro-reactions-radical, Organic-chemistry-3: tut.fi, Retrieved 2, March 2020

8

Polymer Degradation

Any change in the physical or chemical properties of a polymer or polymer-based products due to heat, light, chemicals or any other outer force. There are various processes of polymer degradation like UV degradation, Thermal degradation, thermal depolymerization and photo-oxidation of polymers. The topics elaborated in this chapter will help in gaining a better perspective about degradation of polymers.

Polymer degradation is a change in the properties—tensile strength, color, shape, etc.—of a polymer or polymer-based product under the influence of one or more environmental factors such as heat, light or chemicals such as acids, alkalis and some salts. These changes are usually undesirable, such as cracking and chemical disintegration of products or, more rarely, desirable, as in biodegradation, or deliberately lowering the molecular weight of a polymer for recycling. The changes in properties are often termed "aging".

In a finished product such a change is to be prevented or delayed. Degradation can be useful for recycling/reusing the polymer waste to prevent or reduce environmental pollution. Degradation can also be induced deliberately to assist structure determination.

Polymeric molecules are very large (on the molecular scale), and their unique and useful properties are mainly a result of their size. Any loss in chain length lowers tensile strength and is a primary cause of premature cracking.

Commodity Polymers

Today there are primarily seven commodity polymers in use: polyethylene, polypropylene, polyvinyl chloride, polyethylene terephthalate, polystyrene, polycarbonate, and poly(methyl methacrylate) (Plexiglas). These make up nearly 98% of all polymers and plastics encountered in daily life. Each of these polymers has its own characteristic modes of degradation and resistances to heat, light and chemicals. Polyethylene, polypropylene, and poly(methyl methacrylate) are sensitive to oxidation and UV radiation, while PVC may discolor at high temperatures due to loss of hydrogen chloride gas, and become very brittle. PET is sensitive to hydrolysis and attack by strong acids, while polycarbonate depolymerizes rapidly when exposed to strong alkalis.

For example, polyethylene usually degrades by *random scission*—that is by a random breakage of the linkages (bonds) that hold the atoms of the polymer together. When this polymer is heated above 450 Celsius it becomes a complex mixture of molecules of various sizes that resemble gasoline. Other polymers—like polyalphamethylstyrene—undergo 'specific' chain scission with breakage occurring only at the ends; they literally unzip or depolymerize to become the constituent monomers.

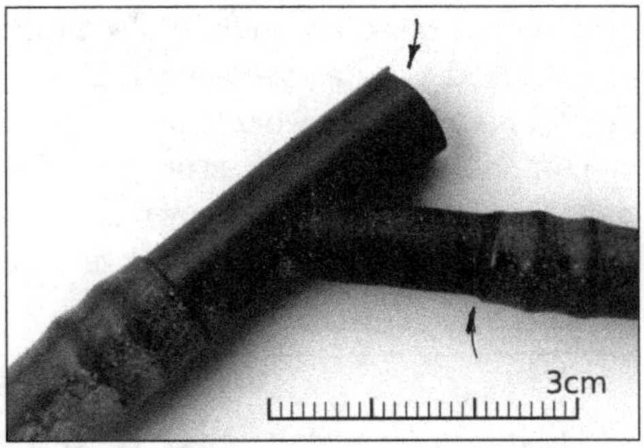

Close-up of broken fuel pipe from road traffic accident.

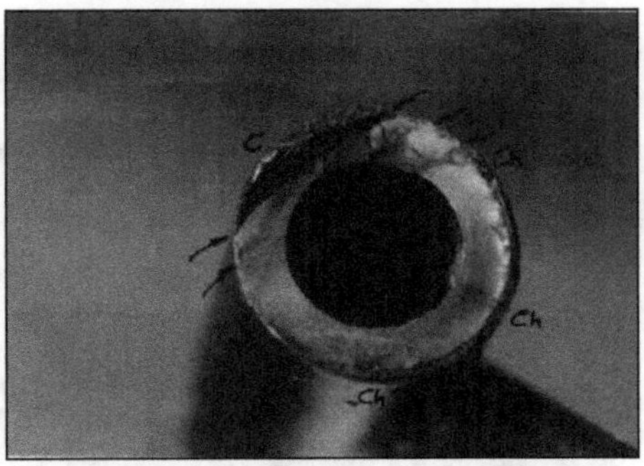

Close-up of broken fuel pipe connector.

Photoinduced Degradation

Most polymers can be degraded by photolysis to give lower molecular weight molecules. Electromagnetic waves with the energy of visible light or higher, such as ultraviolet light, X-rays and gamma rays are usually involved in such reactions.

Thermal Degradation

Chain-growth polymers like poly(methyl methacrylate) can be degraded by thermolysis at high temperatures to give monomers, oils, gases and water. The degradation takes place by:

Thermolysis type	Added material	Temperature	Pressure	Final product
Pyrolysis		Around 500 °C	Reduced pressure	
Hydrogenation	Dihydrogen	Around 450 °C	Around 200 bars	
Gasification	Dioxygen and/or water		Under pressure	Carbon monoxide, Carbon dioxide and hydrogen

Chemical Degradation

Solvolysis

Step-growth polymers like polyesters, polyamides and polycarbonates can be degraded by solvolysis and mainly hydrolysis to give lower molecular weight molecules. The hydrolysis takes place in the presence of water containing an acid or a base as catalyst. Polyamide is sensitive to degradation by acids and polyamide mouldings will crack when attacked by strong acids. For example, the fracture surface of a fuel connector showed the progressive growth of the crack from acid attack (Ch) to the final cusp (C) of polymer. The problem is known as stress corrosion cracking, and in this case was caused by hydrolysis of the polymer. It was the reverse reaction of the synthesis of the polymer:

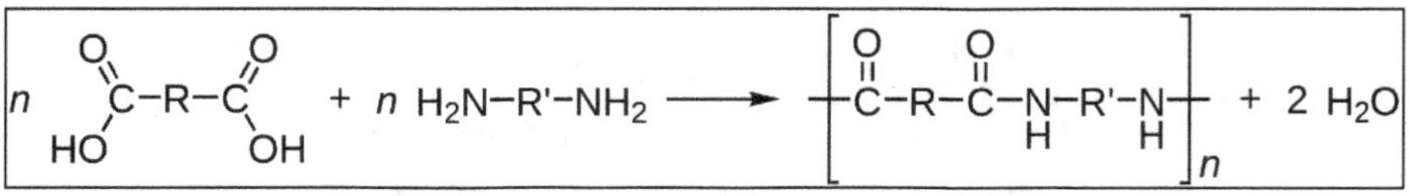

Ozonolysis

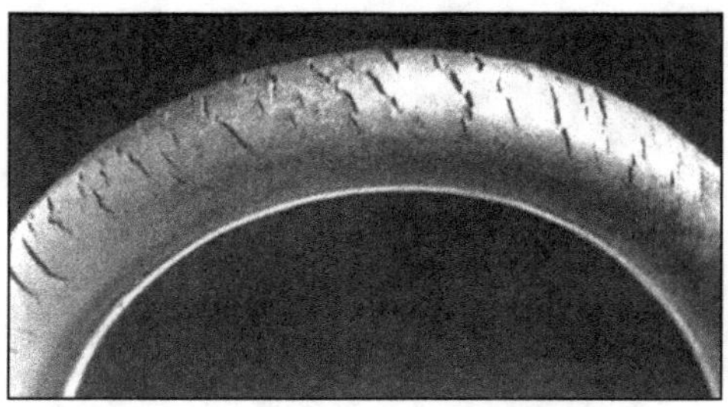

Ozone cracking in Natural rubber tubing.

Cracks can be formed in many different elastomers by ozone attack. Tiny traces of the gas in the air will attack double bonds in rubber chains, with Natural rubber, polybutadiene, Styrene-butadiene rubber and NBR being most sensitive to degradation. Ozone cracks form in products under tension, but the critical strain is very small. The cracks are always oriented at right angles to the strain axis, so will form around the circumference in a rubber tube bent over. Such cracks are dangerous when they occur in fuel pipes because the cracks will grow from the outside exposed surfaces into the bore of the pipe, and fuel leakage and fire may follow. The problem of ozone cracking can be prevented by adding anti-ozonants to the rubber before vulcanization. Ozone cracks were commonly seen in automobile tire sidewalls, but are now seen rarely thanks to these additives. On the other hand, the problem does recur in unprotected products such as rubber tubing and seals.

Oxidation

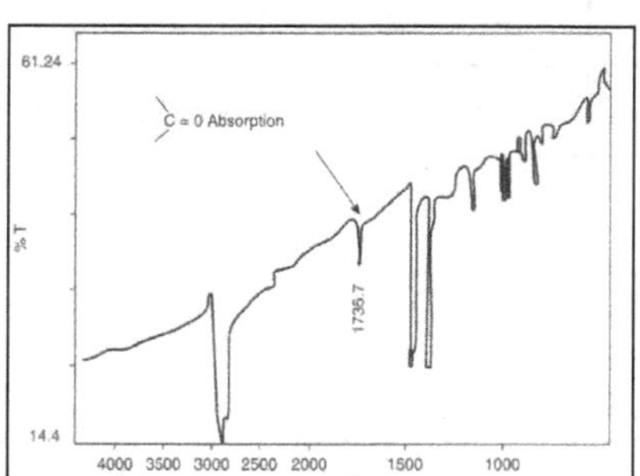

IR spectrum showing carbonyl absorption due to oxidative degradation of polypropylene crutch moulding.

The polymers are susceptible to attack by atmospheric oxygen, especially at elevated temperatures encountered during processing to shape. Many process methods such as extrusion and injection moulding involve pumping molten polymer into tools, and the high temperatures needed for melting may result in oxidation unless precautions are taken. For example, a forearm crutch suddenly snapped and the user was severely injured in the resulting fall. The crutch had fractured across a polypropylene insert within the aluminium tube of the device, and infra-red spectroscopy of the material showed that it had oxidised, possible as a result of poor moulding.

Oxidation is usually relatively easy to detect owing to the strong absorption by the carbonyl group in the spectrum of polyolefins. Polypropylene has a relatively simple spectrum with few peaks at the carbonyl position (like polyethylene). Oxidation tends to start at tertiary carbon atoms because the free radicals formed here are more stable and longer lasting, making them more susceptible to attack by oxygen. The carbonyl group can be further oxidised to break the chain, this weakens the material by lowering its molecular weight, and cracks start to grow in the regions affected.

Galvanic Action

Polymer degradation by galvanic action was first described in the technical literature in 1990. This was the discovery that "plastics can corrode", i.e. polymer degradation may occur through galvanic action similar to that of metals under certain conditions and has been referred to as the "Faudree Effect". In the aerospace field, this finding has largely contributed to aircraft safety, mainly those aircraft that use CFRP and has resulted in a wide body of follow-up research and patents. Normally, when two dissimilar metals such as copper (Cu) and iron (Fe) are put into contact and then immersed in salt water, the iron will undergo corrosion, or rust. This is called a galvanic circuit where the copper is the noble metal and the iron is the active metal, i.e., the copper is the positive (+) electrode and the iron is the negative (-) electrode. A battery is formed. It follows that plastics are made stronger by impregnating them with thin carbon fibers only a few micrometers in diameter known as carbon fiber reinforced polymers (CFRP). This is to produce materials that are high strength and resistant to high temperatures. The carbon fibers act as a noble metal similar to gold (Au) or platinum (Pt). When put into contact with a more active metal, for example with aluminum (Al) in salt water the aluminum corrodes. However, in early 1990, it was reported that imide-linked resins in CFRP composites degrade when bare composite is coupled with an active metal in salt water environments. This is because corrosion not only occurs at the aluminum anode, but also at the carbon fiber cathode in the form of a very strong base with a pH of about 13. This strong base reacts with the polymer chain structure degrading the polymer. Polymers affected include bismaleimides (BMI), condensation polyimides, triazines, and blends thereof. Degradation occurs in the form of dissolved resin and loose fibers. The hydroxyl ions generated at the graphite cathode attack the O-C-N bond in the polyimide structure. Standard corrosion protection procedures were found to prevent polymer degradation under most conditions.

Chlorine-induced Cracking

Another highly reactive gas is chlorine, which will attack susceptible polymers such as acetal resin and polybutylene pipework. There have been many examples of such pipes and acetal fittings failing in properties in the US as a result of chlorine-induced cracking. In essence, the gas attacks sensitive parts of the chain molecules (especially secondary, tertiary, or allylic carbon atoms), oxidizing the chains and ultimately causing chain cleavage. The root cause is traces of chlorine in the water supply, added for its anti-bacterial action, attack occurring even at parts per million traces of the dissolved gas. The chlorine attacks weak parts of a product, and in the case of an acetal resin junction in a water supply system, it is the thread roots that were attacked first, causing a brittle crack to grow. Discoloration on the fracture surface was caused by deposition of carbonates from the hard water supply, so the joint had been in a critical state for many months. The problems in the US also occurred to polybutylene pipework, and led to the material being removed from that market, although it is still used elsewhere in the world.

chlorine attack of acetal resin plumbing joint.

Biological Degradation

Biodegradable plastics can be biologically degraded by microorganisms to give lower molecular weight molecules. To degrade properly biodegradable polymers need to be treated like compost and not just left in a landfill site where degradation is very difficult due to the lack of oxygen and moisture.

Stabilizers

Hindered amine light stabilizers (HALS) stabilize against weathering by scavenging free radicals that are produced by photo-oxidation of the polymer matrix. UV-absorbers stabilizes against weathering by absorbing ultraviolet light and converting it into heat. Antioxidants stabilize the polymer by terminating the chain reaction due to the absorption of UV light from sunlight. The chain reaction initiated by photo-oxidation leads to cessation of crosslinking of the polymers and degradation the property of polymers.

UV Degradation

Many natural and synthetic polymers are attacked by ultraviolet radiation, and products using these materials may crack or disintegrate if they are not UV-stable. The problem is known as *UV degradation*, and is a common problem in products exposed to sunlight. Continuous exposure is a more serious problem than intermittent exposure, since attack is dependent on the extent and degree of exposure.

Many pigments and dyes can also be affected, and the problem known as phototendering can affect textiles such as curtains or drapes.

Susceptible Polymers

Effect of UV exposure on polypropylene rope.

Common synthetic polymers that can be attacked include polypropylene and LDPE, where tertiary carbon bonds in their chain structures are the centres of attack. Ultraviolet rays interact with these bonds to form free radicals, which then react further with oxygen in the atmosphere, producing carbonyl groups in the main chain. The exposed surfaces of products may then discolour and crack, and in extreme cases, complete product disintegration can occur.

In fibre products like rope used in outdoor applications, product life will be low because the outer fibres will be attacked first, and will easily be damaged by abrasion for example. Discolouration of the rope may also occur, thus giving an early warning of the problem.

Polymers which possess UV-absorbing groups such as aromatic rings may also be sensitive to UV degradation. Aramid fibres like Kevlar, for example, are highly UV-sensitive and must be protected from the deleterious effects of sunlight.

Detection

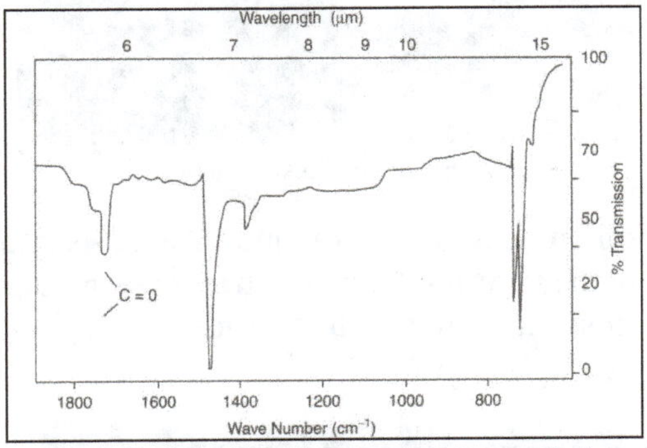

IR spectrum showing carbonyl absorption due to UV degradation of polyethylene.

The problem can be detected before serious cracks are seen in a product using infrared spectroscopy, where attack occurs by oxidation of bonds activated by the UV radiation forming carbonyl groups in the polymer chains.

In the example shown at left, carbonyl groups were easily detected by IR spectroscopy from a cast thin film. The product was a road cone made by rotational moulding in LDPE, which had cracked prematurely in service. Many similar cones also failed because an anti-UV additive had not been used during processing. Other plastic products which failed included polypropylene mancabs used at roadworks which cracked after service of only a few months.

Prevention

UV attack by sunlight can be ameliorated or prevented by adding anti-UV chemicals to the polymer when mixing the ingredients, prior to shaping the product by injection moulding for example. UV stabilizers in plastics usually act by absorbing the UV radiation preferentially, and dissipating the energy as low-level heat. The chemicals used are similar to those in sunscreen products, which protect skin from UV attack. Frequently, glass can be a better alternative to polymers when it comes to UV degradation. Most of the commonly used glass types are highly resistant to UV radiation. Explosion protection lamps for oil rigs for example can be made either from polymer or glass. Here, the UV radiation and rough weathers belabor the polymer so much, that the material has to be replaced frequently.

Materials Testing

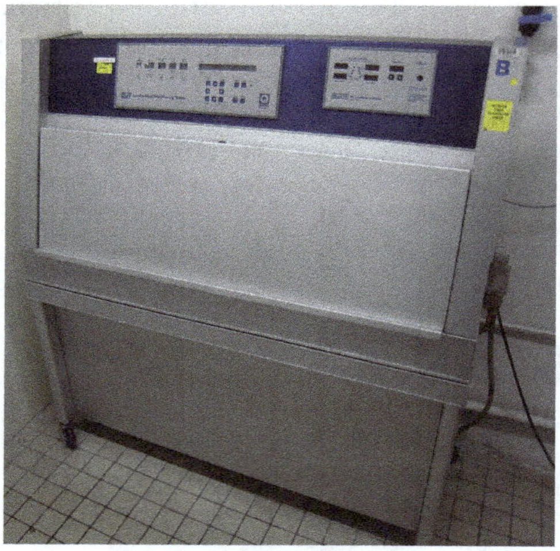

Example of test device.

The effects of UV degradation on materials that require a long service life can be measured with accelerated exposure tests. With modern solar concentrator technologies, it is possible to simulate 63 years of natural UV radiation exposure on a test device in a single year.

Thermal Degradation of Polymers

Thermal degradation of polymers is molecular deterioration as a result of overheating. At high temperatures the components of the long chain backbone of the polymer can begin to separate (molecular scission) and react with one another to change the properties of the polymer. Thermal degradation can present an upper limit to the service temperature of plastics as much as the possibility of mechanical property loss. Indeed unless correctly prevented, significant thermal degradation can occur at temperatures much lower than those at which mechanical failure is likely to occur. The chemical reactions involved in thermal degradation lead to physical and optical property changes relative to the initially specified properties. Thermal degradation generally involves changes to the molecular weight (and molecular weight distribution) of the polymer and typical property changes include reduced ductility and embrittlement, chalking, color changes, cracking, general reduction in most other desirable physical properties. Thermal breakdown products may include a complex mixture of compounds, including but not limited to carbon monoxide, ammonia, aliphatic amines, ketones, nitriles, and hydrogen cyanide, which may be flammable, toxic and/or irritating. The specific materials generated will vary depending on the additives and colorants used, specific temperature, time of exposure and other immediate environmental factors.

The Mechanism of Thermal Degradation

Most types of degradation follow a similar basic pattern. The conventional model for thermal degradation is that of an auto-oxidation process which involves the major steps of initiation, propagation, branching, and termination.

Initiation

The initiation of thermal degradation involves the loss of a hydrogen atom from the polymer chain as a result of energy input from heat or light. This creates a highly reactive and unstable polymer 'free radical' (R•) and a hydrogen atom with an unpaired electron (H•).

Propagation

The propagation of thermal degradation can involve a variety of reactions and one of these is where the free radical (R•) reacts with an oxygen (O_2) molecule to form a peroxy radical (ROO•) which can then remove a hydrogen atom from another polymer chain to form a hydroperoxide (ROOH) and so regenerate the free radical (R•). The hydroperoxide can then split into two new free radicals, (RO•) + (•OH), which will continue to propagate the reaction to other polymer molecules. The process can therefore accelerate depending on how easy it is to remove the hydrogen from the polymer chain.

Termination

The termination of thermal degradation is achieved by 'mopping up' the free radicals to create inert products. This can occur naturally by combining free radicals or it can be assisted by using stabilizers in the plastic.

The Research Methods of Thermal Degradation of Polymers

TGA

(Thermogravimetric analysis) (TGA) refers to the techniques where a sample is heated in a controlled atmosphere at a defined heating rate whilst the samples mass is measured. When a polymer sample degrades, its mass decreases due to the production of gaseous products like carbon monoxide, water vapour and carbon dioxide.

DTA and DSC

(Differential thermal analysis) (DTA) and (differential scanning calorimetry) (DSC): Analyzing the heating effect of polymer during the physical changes in terms of glass transition, melting, and so on. These techniques measure the heat flow associated with oxidation.

Ways of Polymer Thermal Degradation

Depolymerisation

Under thermal effect, the end of polymer chain departs, and forms low free radical which has low activity. Then according to the chain reaction mechanism, the polymer loses the monomer one by one. However, the molecular chain doesn't change a lot in a short time. The reaction is shown below. This process is common for polymethymethacrylate (perspex).

$CH_2\text{-}C(CH_3)COOCH_3\text{-}CH_2\text{-}C^*(CH_3)COOCH_3 \rightarrow CH_2\text{-}C^*(CH_3)COOCH_3 + CH_2{=}C(CH_3)COOCH_3$

Random Chain Scission

The backbone will break down randomly, this can occur at any position of the backbone, as a result the molecular weight decreases rapidly. As new free radicals with high reactivity are formed, monomers cannot be a product of this reaction, also intermolecular chain transfer and disproportion termination reactions can occur.

$CH_2\text{-}CH_2\text{-}CH_2\text{-}CH_2\text{-}CH_2\text{-}CH_2\text{-}CH_2' \rightarrow CH_2\text{-}CH_2\text{-}CH{=}CH_2 + CH_3\text{-}CH_2\text{-}CH_2'$ or $CH_2' + CH_2{=}CH\text{-}CH_2\text{-}CH_2\text{-}CH_2\text{-}CH_3$

Side-group Elimination

Groups that are attached to the side of the backbone are held by bonds which are weaker than the bonds connecting the chain. When the polymer is heated, the side groups are stripped off from the chain before it is broken into smaller pieces. For example the PVC eliminates HCl, under 100–120 °C.

Thermal Depolymerization

Thermal depolymerization (TDP) is a depolymerization process using hydrous pyrolysis for the reduction of complex organic materials (usually waste products of various sorts, often biomass and plastic) into light crude oil. It mimics the natural geological processes thought to be involved in the production of fossil fuels. Under pressure and heat, long chain polymers of hydrogen, oxygen, and carbon decompose into short-chain petroleum hydrocarbons with a maximum length of around 18 carbons.

Similar Processes

Thermal depolymerisation is similar to other processes which use superheated water as a major step to produce fuels, such as direct Hydrothermal Liquefaction. These are distinct from processes using dry materials to depolymerize, such as pyrolysis. The term Thermochemical Conversion (TCC) has also been used for conversion of biomass to oils, using superheated water, although it is more usually applied to fuel production via pyrolysis. Other commercial scale processes include the "SlurryCarb" process operated by EnerTech, which uses similar technology to decarboxylate wet solid biowaste, which can then be physically dewatered and used as a solid fuel called E-Fuel. The plant was designed to Rialto to process 683 tons of waste per day. However, it failed to perform to design standards and was closed down. The Rialto facility defaulted on its bond payments and is in the process of being liquidated. The Hydro Thermal Upgrading (HTU) process uses superheated water to produce oil from domestic waste. A demonstration plant is due to start up in The Netherlands said to be capable of processing 64 tons of biomass (dry basis) per day into oil. Thermal depolymerisation differs in that it contains a hydrous process followed by an anhydrous cracking / distillation process.

History

Thermal depolymerization is similar to the geological processes that produced the fossil fuels used today, except that the technological process occurs in a timeframe measured in hours. Until recently, the human-designed processes were not efficient enough to serve as a practical source of fuel—more energy was required than was produced.

The first industrial process to obtain gas, diesel fuels and other petroleum products through pyrolysis of coal, tar or biomass was designed and patented in the late 1920s by Fischer-Tropsch. In U. S. patent 2,177,557, issued in 1939, Bergstrom and Cederquist discuss a method for obtaining oil from wood in which the wood is heated under pressure in water with a significant amount of calcium hydroxide added to the mixture. In the early 1970s Herbert R. Appell and coworkers worked with hydrous pyrolysis methods, as exemplified by U. S. patent 3,733,255 (issued in 1973), which discusses the production of oil from sewer sludge and municipal refuse by heating the material in water, under pressure, and in the presence of carbon monoxide.

An approach that exceeded break-even was developed by Illinois microbiologist Paul Baskis in the 1980s and refined over the next 15 years (see U. S. patent 5,269,947, issued in 1993). The technology was finally developed for commercial use in 1996 by Changing World Technologies (CWT). Brian S. Appel (CEO of CWT) took the technology in 2001 and expanded and changed it into what is now referred to as TCP (Thermal Conversion Process), and has applied for and obtained several patents. A Thermal Depolymerization demonstration plant was completed in 1999 in Philadelphia by Thermal Depolymerization, LLC, and the first full-scale commercial plant was constructed in Carthage, Missouri, about 100 yards (91 m) from ConAgra Foods' massive Butterball turkey plant, where it is expected to process about 200 tons of turkey waste into 500 barrels (79 m^3) of oil per day.

Theory and Process

In the method used by CWT, the water improves the heating process and contributes hydrogen to the reactions.

In the Changing World Technologies (CWT) process, the feedstock material is first ground into small chunks, and mixed with water if it is especially dry. It is then fed into a pressure vessel reaction chamber where it is heated at constant volume to around 250 °C. Similar to a pressure cooker (except at much higher pressure), steam naturally raises the pressure to 600 psi (4 MPa) (near the point of saturated water). These conditions are held for approximately 15 minutes to fully heat the mixture, after which the pressure is rapidly released to boil off most of the water. The result is a mix of crude hydrocarbons and solid minerals. The minerals are removed, and the hydrocarbons are sent to a second-stage reactor where they are heated to 500 °C, further breaking down the longer hydrocarbon chains. The hydrocarbons are then sorted by fractional distillation, in a process similar to conventional oil refining.

The CWT company claims that 15 to 20% of feedstock energy is used to provide energy for the plant. The remaining energy is available in the converted product. Working with turkey offal as the feedstock, the process proved to have yield efficiencies of approximately 85%; in other words, the energy contained in the end products of the process is 85% of the energy contained in the inputs to the process (most notably the energy content of the feedstock, but also including electricity for pumps and natural gas or woodgas for heating). If one considers the energy content of the feedstock to be free (i.e., waste material from some other process), then 85 units of energy are made available for every 15 units of energy consumed in process heat and electricity. This means the "Energy Returned on Energy Invested" (EROEI) is (6.67), which is comparable to other energy harvesting processes. Higher efficiencies may be possible with drier and more carbon-rich feedstocks, such as waste plastic.

By comparison, the current processes used to produce ethanol and biodiesel from agricultural sources have EROEI in the 4.2 range, when the energy used to produce the feedstocks is accounted for (in this case, usually sugar cane, corn, soybeans and the like). These EROEI values are not directly comparable, because these EROEI calculations include the energy cost to produce the feedstock, whereas the above EROEI calculation for thermal depolymerization process (TDP) does not.

The process breaks down almost all materials that are fed into it. TDP even efficiently breaks down many types of hazardous materials, such as poisons and difficult-to-destroy biological agents such as prions.

Feedstocks and outputs with thermal depolymerization

Average TDP Feedstock Outputs				
Feedstock	**Oils**	**Gases**	**Solids (mostly carbon based)**	**Water (Steam)**
Plastic bottles	70%	16%	6%	8%
Medical waste	65%	10%	5%	20%
Tires	44%	10%	42%	4%
Turkey offal	39%	6%	5%	50%
Sewage sludge	26%	9%	8%	57%
Paper (cellulose)	8%	48%	24%	20%

(Note: Paper/cellulose contains at least 1% minerals, which was probably grouped under carbon solids.)

Carthage Plant Products

As reported on 04/02/2006 by Discover Magazine, a Carthage, Missouri plant was producing 500 barrels per day (79 m^3/d) of oil made from 270 tons of turkey entrails and 20 tons of hog lard. This represents an oil yield of 22.3 percent. The Carthage plant produces API 40+, a high value crude oil. It contains light and heavy naphthas, a kerosene, and a gas oil fraction, with essentially no heavy fuel oils, tars, asphaltenes or waxes. It can be further refined to produce No. 2 and No. 4 fuel oils.

TDP-40 Oil Classification by D-5443 PONA method	
Output Material	**% by Weight**
Paraffins	22%
Olefins	14%
Naphthenes	3%
Aromatics	6%
C14/C14+	55%
	100%

The fixed carbon solids produced by the TDP process have multiple uses as a filter, a fuel source and a fertilizer. It can be used as activated carbon in wastewater treatment, as a fertilizer, or as a fuel similar to coal.

Advantages

The process can break down organic poisons, due to breaking chemical bonds and destroying the molecular shape needed for the poison's activity. It is likely to be highly effective at killing pathogens, including prions. It can also safely remove heavy metals from the samples by converting them from their ionized or organometallic forms to their stable oxides which can be safely separated from the other products.

Along with similar processes, it is a method of recycling the energy content of organic materials without first removing the water. It can produce liquid fuel, which separates from the water physically without need for drying. Other methods to recover energy often require pre-drying (e.g. burning, pyrolysis) or produce gaseous products (e.g. anaerobic digestion).

Potential Sources of Waste Inputs

The United States Environmental Protection Agency estimates that in 2006 there were 251 million tons of municipal solid waste, or 4.6 pounds generated per day per person in the USA. Much of this mass is considered unsuitable for oil conversion.

Limitations

The process only breaks long molecular chains into shorter ones, so small molecules such as carbon dioxide or methane cannot be converted to oil through this process. However, the methane in the feedstock is recovered and burned to heat the water that is an essential part of the process. In addition, the gas can be burned in a combined heat and power plant, consisting of a gas turbine which drives a generator to create electricity, and a heat exchanger to heat the process input water from the exhaust gas. The electricity can be sold to the power grid, for example under a feed-in tariff scheme. This also increases the overall efficiency of the process (already said to be over 85% of feedstock energy content).

Another option is to sell the methane product as biogas. For example, biogas can be compressed, much like natural gas, and used to power motor vehicles.

Many agricultural and animal wastes could be processed, but many of these are already used as fertilizer, animal feed, and, in some cases, as feedstocks for paper mills or as boiler fuel. Energy crops constitute another potentially large feedstock for thermal depolymerization.

Current Status

Reports in 2004 claimed that the Carthage facility was selling products at 10% below the price of equivalent oil, but its production costs were low enough that it produced a profit. At the time it was paying for turkey waste.

The plant then consumed 270 tons of turkey offal (the full output of the turkey processing plant) and 20 tons of egg production waste daily. In February 2005, the Carthage plant was producing about 400 barrels per day (64 m^3/d) of crude oil.

In April 2005 the plant was reported to be running at a loss. Further 2005 reports summarized some economic setbacks which the Carthage plant encountered since its planning stages. It was thought that concern over mad cow disease would prevent the use of turkey waste and other animal products as cattle feed, and thus this waste would be free. As it turned out, turkey waste may still be used as feed in the United States, so that the facility must purchase that feed stock at a cost of $30 to $40 per ton, adding $15 to $20 per barrel to the cost of the oil. Final cost, as of January 2005, was $80/barrel ($1.90/gal).

The above cost of production also excludes the operating cost of the thermal oxidizer and scrubber added in May 2005 in response to odor complaints.

A biofuel tax credit of roughly $1 per US gallon (26 ¢/L) on production costs was not available because the oil produced did not meet the definition of "biodiesel" according to the relevant American tax legislation. The Energy Policy Act of 2005 specifically added thermal depolymerization to a $1 renewable diesel credit, which became effective at the end of 2005, allowing a profit of $4/barrel of output oil.

Company Expansion

The company has explored expansion in California, Pennsylvania, and Virginia, and is presently examining projects in Europe, where animal products cannot be used as cattle feed. TDP is also being considered as an alternative means for sewage treatment in the United States.

Smell Complaints

The pilot plant in Carthage was temporarily shut down due to smell complaints. It was soon re-started when it was discovered that few of the odors were generated by the plant. Furthermore, the plant agreed to install an enhanced thermal oxidizer and to upgrade its air scrubber system under a court order. Since the plant is located only four blocks from the tourist-attracting town center, this has strained relations with the mayor and citizens of Carthage.

According to a company spokeswoman, the plant has received complaints even on days when it is not operating. She also contended that the odors may not have been produced by their facility, which is located near several other agricultural processing plants.

On December 29, 2005, the plant was ordered by the state governor to shut down once again over allegations of foul odors as reported by MSNBC.

As of March 7, 2006, the plant has begun limited test runs to validate it has resolved the odor issue.

As of August 24, 2006, the last lawsuit connected with the odor issue has been dismissed and the problem is acknowledged as fixed. In late November, however, another complaint was filed over bad smells. This complaint was closed on January 11 of 2007 with no fines assessed.

Status as of February 2009

A May 2003 article in Discover magazine stated, "Appel has lined up federal grant money to help build demonstration plants to process chicken offal and manure in Alabama and crop residuals and grease in Nevada. Also in the works are plants to process turkey waste and manure in Colorado and pork and cheese waste in Italy. He says the first generation of depolymerization centers will be up and running in 2005. By then it should be clear whether the technology is as miraculous as its backers claim."

However, as of August 2008, the only operational plant listed at the company's website is the initial one in Carthage, Missouri.

Changing World Technology applied for an IPO on August 12; 2008, hoping to raise $100 million.

The unusual Dutch Auction type IPO failed possibly because CWT has lost nearly $20 million with very little revenue.

CWT, the parent company of Renewable Energy Solutions, filed for Chapter 11 bankruptcy. No details on plans for the Carthage plant have been released.

In April 2013, CWT was acquired by a Canadian firm, Ridgeline Energy Services, based in Calgary.

Similar Technologies

- Plasma Converters use powerful electric arcs to reduce and extract energy from waste.
- Wet oxidation
- Hydrocracking

Chemically Assisted Degradation of Polymers

Chemically assisted degradation of polymers is a type of polymer degradation that involves a change of the polymer properties due to a chemical reaction with the polymer's surroundings. There are many different types of possible chemical reactions causing degradation however most of these reactions result in the breaking of double bonds within the polymer structure.

Examples of Chemically Assisted Degradation

Degradation of Rubber by Ozone

One common example of chemically assisted degradation is the degradation of rubber by ozone particles. Ozone is a naturally occurring atmospheric molecule that is produced by electric discharge or through a reaction of Oxygen with solar radiation. Ozone is also produced with atmospheric pollutants reacted with ultraviolet radiation. For a reaction to occur, ozone concentrations only have to be as low as 3-5 parts per hundred million (pphm) and when these concentrations are reached, a reaction occurs with a thin surface layer (5 x10-7 metres) of the material. The ozone molecules react with the rubber which in most cases is unsaturated (contains double bonds), however a reaction will still occur in saturated polymers (those containing only single bonds). When reaction occurs, scission of the polymer chain (breaking of double covalent bonds) takes place forming decomposition products:

Chain scission increases with the presence of active Hydrogen molecules (for example, in water) as well as acids and alcohols. Along with this type of reaction, cross linking and side branch formations also occur by an activation of the double bond and these make the rubber material more brittle. Due to the increase in brittleness due to the chemical reactions, cracks form in areas of high stress. As propagation of these cracks increases, new surfaces are opened for degradation to occur.

Degradation of Poly(Vinyl) Chloride (PVC)

Degradation can also occur as a result of the formation, and then breakage of double bonds, such as solvolysis in PVC(Peacock). Solvolysis occurs when a Carbon-X bond, with X representing a halogen, is broken. This occurs in PVC in the presence of an acid species. Active Hydrogen atoms will remove a Chlorine atom from the polymer molecule, forming Hydrochloric acid (HCl). The HCl produced may then cause dechlorination of adjacent Carbon atoms. The dechlorinated Carbon atoms then tend to form double bonds, which can be attacked and broken by ozone, just like the degradation of rubbers described above.

Degradation of Polyester

The degradation of a polyester may occur without the presence of the acidic catalyst that causes degradation of PVC. During hydrolysis water acts as the reactive catalyst instead of the acid. It causes degradation mainly at high temperature and pressure during processing.

In this process the water molecule will attack the C-O ester bond, splitting the polymer in half. The water molecule will then dissociate, with one Hydrogen atom forming a carboxylic acid group on the Carbon atom with the double bonded Oxygen, while the remaining atoms form an alcohol on the other chain end. These reactive products may also cause further degradation of the polymer chain. This chain scission lowers average molecular weight of the polymer, decreasing the number and strength of intermolecular bonds as well as the degree of entanglement. This will increase chain mobility, decreasing strength of the polymer and increasing deformation at low stresses.

Protection Against Chemically Assisted Degradation

Both physical and chemical barriers can be used to protect a polymer from chemically assisted degradation. A physical barrier must provide continuous protection, must not react with the polymer's environment, must be flexible so that stretching may occur and must also be able to regenerate (after wear processes). A chemical barrier must be highly reactive with the polymer's surroundings so that the barrier reacts with the environmental conditions rather than the polymer itself. This barrier involves addition of a material into the polymer blend during fabrication of the polymer. Due to this, the barrier addition must have a suitable solubility, must be economically feasible and must not hinder the production process. For the barrier to be activated, the addition must diffuse to the surface and so a suitable diffusivity is also required. There are four theories on how these types of barriers protect the polymer material:

- *Scavenger theory*: the protective layer reacts with the ozone rather than the polymer.
- *Protective film theory*: the protective layer reacts with the polymer producing a thin film on the polymer surface which is inert and can't be penetrated.
- *Re-linking theory*: the protective layer causes broken double bonds to be reformed.
- *Self-healing theory*: the protective layer reacts with degraded polymer chains to form low-molecular-weight material which forms an inert film on the surface.

Of these theories, the scavenger theory is the most common and most important. However, more than one theory can act at the same time and the theory that takes place depends on the protective materials, the polymer and surrounding environment.

Photo-oxidation of Polymers

Photo-oxidation is the degradation of a polymer surface in the presence of oxygen or ozone. The effect is facilitated by radiant energy such as UV or artificial light. This process is the most significant factor in weathering of polymers. Photo-oxidation is a chemical change that reduces the polymer's molecular weight. As a consequence of this change the material becomes more brittle, with a reduction in its tensile, impact and elongation strength. Discoloration and loss of surface smoothness accompany photo-oxidation. High temperature and localized stress concentrations are factors that significantly increase the effect of photo-oxidation.

Photo-oxidation Protection

Poly(ethylene-naphthalate) (PEN) can be protected by applying a zinc oxide coating, which acts as protective film reducing the diffusion of oxygen. Zinc oxide can also be used on polycarbonate (PC) to decrease the oxidation and photo-yellowing rate caused by solar radiation.

Effects of Dyes/Pigments

Adding pigment light absorbers and photostabilizers (UV absorbers) is one way to minimise photo-oxidation in polymers. Antioxidants are used to inhibit the formation of hydroperoxides in the photo-oxidation process.

Dyes and pigments are used in polymer materials to provide color changing properties. These additives can reduce the rate of polymer degradation. Cu-phthalocyanine dye can help stabilize against degradation, but in other situations such as photochemical aging can actually accelerate degradation. The excited Cu-phthalocyanine may abstract hydrogen atoms from methyl groups in the PC, which increase the formation of free radicals. This acts as the starting points for the sequential photo-oxidation reactions leading to the degradation of the PC.

Electron transfer sensitization is a mechanism where the excited Cu-phthalocyanine abstracts electrons from PC to form Cu-Ph radical anion and PC radical cations. These species in the presence of oxygen can cause oxidation of the aromatic ring.

Chemical Mechanism

Aldehydes, ketones and carboxylic acids along or at the end of polymer chains are generated by oxygenated species in photolysis of photo-oxidation. The initiation of photo-oxidation reactions is due to the existence of chromophoric groups in the macromolecules. Photo-oxidation can occur simultaneously with thermal degradation and each of these effects can accelerate the other.

$Polymer \longrightarrow P\bullet + P\bullet$	Initial step
$P\bullet + O_2 \longrightarrow POO\bullet$ $POO\bullet + PH \longrightarrow POOH + P\bullet$	Chain Propagation
$POOH \longrightarrow PO\bullet + \bullet OH$ $PH + \bullet OH \longrightarrow P\bullet + H_2O$ $PO\bullet \longrightarrow$ Chain scisson reactions	Chain branching
$POO\bullet + POO\bullet$ $POO\bullet + P\bullet$ $P\bullet + P\bullet$ $\longrightarrow$ Cross linking reactions to non-radical products	Termination

where PH = Polymer

P• = Polymer alkyl radical

PO• = Polymer oxy radical (Polymer alkoxy radical)

POO• = Polymer peroxy radical (Polymer alkylperoxy radical)

POOH = Polymer hydroperoxide

HO• = hydroxy radical

The photo-oxidation reactions include chain scission, cross linking and secondary oxidative reactions. The following process steps can be considered:

1. Initial step: Free radicals are formed by photon absorption.
2. Chain Propagation step: A free radical reacts with oxygen to produce a polymer peroxy radical (POO•). This reacts with a polymer molecule to generate polymer hydroperoxide (POOH) and a new polymer alkyl radical (P•).
3. Chain Branching: Polymer oxy radicals (PO•) and hydroxy radicals (HO•) are formed by photolysis.
4. Termination step: Cross linking is a result of the reaction of different free radicals with each other.

References

- "THE PHOTO-OXIDATION OF POLYMERS - A comparison with low molecular weight compounds" (PDF). Pergamon Press Ltd. 1979 - Pure & Appi. Chem., Vol. 51, pp.233—240. Retrieved 20, February 2020
- Faudree, Michael C. (1991). "Relationship of Graphite/Polyimide Composites to Galvanic Processes" (PDF). Society for the Advancement of Material and Process Engineering (SAMPE) Journal. 2: 1288–1301. ISBN 0-938994-56-5
- "Photo-oxidisation of electroluminescent polymers studied by core level photoabsorption specttroscopy" (PDF). American institute of physics 1996. Retrieved 9, February 2020

Permissions

All chapters in this book are published with permission under the Creative Commons Attribution Share Alike License or equivalent. Every chapter published in this book has been scrutinized by our experts. Their significance has been extensively debated. The topics covered herein carry significant information for a comprehensive understanding. They may even be implemented as practical applications or may be referred to as a beginning point for further studies.

We would like to thank the editorial team for lending their expertise to make the book truly unique. They have played a crucial role in the development of this book. Without their invaluable contributions this book wouldn't have been possible. They have made vital efforts to compile up to date information on the varied aspects of this subject to make this book a valuable addition to the collection of many professionals and students.

This book was conceptualized with the vision of imparting up-to-date and integrated information in this field. To ensure the same, a matchless editorial board was set up. Every individual on the board went through rigorous rounds of assessment to prove their worth. After which they invested a large part of their time researching and compiling the most relevant data for our readers.

The editorial board has been involved in producing this book since its inception. They have spent rigorous hours researching and exploring the diverse topics which have resulted in the successful publishing of this book. They have passed on their knowledge of decades through this book. To expedite this challenging task, the publisher supported the team at every step. A small team of assistant editors was also appointed to further simplify the editing procedure and attain best results for the readers.

Apart from the editorial board, the designing team has also invested a significant amount of their time in understanding the subject and creating the most relevant covers. They scrutinized every image to scout for the most suitable representation of the subject and create an appropriate cover for the book.

The publishing team has been an ardent support to the editorial, designing and production team. Their endless efforts to recruit the best for this project, has resulted in the accomplishment of this book. They are a veteran in the field of academics and their pool of knowledge is as vast as their experience in printing. Their expertise and guidance has proved useful at every step. Their uncompromising quality standards have made this book an exceptional effort. Their encouragement from time to time has been an inspiration for everyone.

The publisher and the editorial board hope that this book will prove to be a valuable piece of knowledge for students, practitioners and scholars across the globe.

Index